Lecture Notes in Mechanical Engineering

Lecture Notes in Mechanical Engineering (LNME) publishes the latest developments in Mechanical Engineering—quickly, informally and with high quality. Original research or contributions reported in proceedings and post-proceedings represents the core of LNME. Volumes published in LNME embrace all aspects, subfields and new challenges of mechanical engineering.

To submit a proposal or request further information, please contact the Springer Editor of your location:

Europe, USA, Africa: Leontina Di Cecco at Leontina.dicecco@springer.com
China: Ella Zhang at ella.zhang@springer.com
India: Priya Vyas at priya.vyas@springer.com
Rest of Asia, Australia, New Zealand: Swati Meherishi at swati.meherishi@springer.com

Topics in the series include:

- Engineering Design
- Machinery and Machine Elements
- Mechanical Structures and Stress Analysis
- Automotive Engineering
- Engine Technology
- Aerospace Technology and Astronautics
- Nanotechnology and Microengineering
- Control, Robotics, Mechatronics
- MEMS
- Theoretical and Applied Mechanics
- Dynamical Systems, Control
- Fluid Mechanics
- Engineering Thermodynamics, Heat and Mass Transfer
- Manufacturing Engineering and Smart Manufacturing
- Precision Engineering, Instrumentation, Measurement
- Materials Engineering
- Tribology and Surface Technology

Indexed by SCOPUS, EI Compendex, and INSPEC.

All books published in the series are evaluated by Web of Science for the Conference Proceedings Citation Index (CPCI).

To submit a proposal for a monograph, please check our Springer Tracts in Mechanical Engineering at https://link.springer.com/bookseries/11693.

Achim Wagner · Kosmas Alexopoulos ·
Sotiris Makris
Editors

Advances in Artificial Intelligence in Manufacturing

Proceedings of the 1st European Symposium on Artificial Intelligence in Manufacturing, September 19, 2023, Kaiserslautern, Germany

Springer

Editors
Achim Wagner
German Research Center for Artificial Intelligence
Kaiserslautern, Germany

Sotiris Makris
Laboratory for Manufacturing Systems & Automation
University of Patras
Patras, Greece

Kosmas Alexopoulos
Laboratory for Manufacturing Systems & Automation
University of Patras
Patras, Greece

ISSN 2195-4356 ISSN 2195-4364 (electronic)
Lecture Notes in Mechanical Engineering
ISBN 978-3-031-57495-5 ISBN 978-3-031-57496-2 (eBook)
https://doi.org/10.1007/978-3-031-57496-2

This Springer imprint is published by the registered company Springer Nature Switzerland AG
The registered company address is: Gewerbestrasse 11, 6330 Cham, Switzerland

Preface

This volume of Lecture Notes in Mechanical Engineering contains selected papers presented at the 1st European Symposium on Artificial Intelligence in Manufacturing (ESAIM 2023), held in Kaiserslautern, Germany, on September 19, 2023. The conference was organized by the German Research Center for Artificial Intelligence (DKFI) and the Laboratory for Manufacturing Systems & Automation (LMS) of the University of Patras, Greece.

Artificial intelligence (AI) technology becomes mature, and at the same time researchers and manufacturers discover new applications in which AI can support manufacturing operations. The real benefit from AI in manufacturing will not just be by automating tasks but provide new levels of autonomy and human assistance that will make possible entirely new applications and introduce new business processes in manufacturing. In contrast to consumer applications, well known from mobile device, media and automotive domains, industrial AI applications in general cannot refer to a big data availability collected during operation and interaction with the users. Additional challenges arise related to the industrial conditions such as privacy, safety, interoperability and physics-induced real-time requirements. Due to the lack of a huge amount of centralized data and the requirements on the shop floor, new approaches for data collection, processing and modeling are required, which are reflected by various AI methods and technologies applied to digitized factories.

The scope of the ESAIM'23 symposium was around recent developments of AI in manufacturing, advancing key concepts and technologies, as well as understanding the benefits and the barriers when applying such technologies in industrial practice. The symposium has welcomed contributions focusing on theoretical, applied research and industrial case studies.

ESAIM'23 covered topics such as AI in manufacturing processes, robots, machines and operations support in manufacturing and AI in manufacturing systems. It also considered research work on cross-cutting aspects such as information systems, regulation, education, systems engineering and data augmentation.

ESAIM'23 received 26 contributions from organizations and researchers in Europe. After a thorough peer-review process, the Program Committee accepted 19 papers. Thank you very much to the authors for their contribution. These papers are published in the present book, achieving an acceptance rate of about 73%.

We would like to thank members of the Program Committee and invited external reviewers for their efforts and expertise in contributing to reviewing, without which it would be impossible to maintain the high standards of peer-reviewed papers.

Many thanks to keynote speaker Prof. Martin Ruskowski, Department of Machine Tools and Control Systems (WSKL), University of Kaiserslautern-Landau for sharing his knowledge and experience in the topic of AI in manufacturing.

The book “Advances in Artificial Intelligence in Manufacturing” consists of three parts. The first part aims at recent developments in “Artificial Intelligence at Manufacturing System Level”. A manufacturing system can be considered a combination of machines, cells, intra-logistics devices and other peripheral devices, used on the factory floor as well as in logistics. A range of topics such as AI for production planning and scheduling, AI for condition monitoring and digital platforms and frameworks for AI were discussed. Active learning frameworks for the deployment of machine learning development can be used for learning efficiently from limited labeled data, allowing manufacturers to optimize production processes with better accuracy and reduced data annotation efforts. Machine learning algorithms for fusing and making sense of big multichannel data gathered in discrete manufacturing applications to anticipate the detection of anomalies are also being discussed. In a similar manner Asset Administration Shell (AAS) and Industrial Data Spaces (IDS) can be used for setting up architectures and supporting interoperability for AI applications in manufacturing value chains. Going more into shop-floor applications, skill-based production by means of incorporating automated fault detection can be used for modeling the behavior of Cyber-Physical Production Modules. Runge-Kutta Neural Networks (RKNN) are used to detect rare fault cases for mobile robot applications, while Physics-Informed Neural Networks (PINNs) can be used in parameter identification for continuum models of manufacturing systems and deep reinforcement learning agents can be trained for optimal dispatch rules selection in production scheduling.

The second part is based on recent developments in the field “Artificial Intelligence at Manufacturing Equipment Level”. There is a wide range of topics addressed in this chapter such as AI for flexible and precise robotics, AI for exoskeletons, AI for enhanced human-robot collaboration, AI for quality assessment and defects detection. In the field of quality assessment, unsupervised machine learning is investigated for blind rivet inspection and a set of tools for defect detection tasks within a factory, based on deep learning methods as elements of different quality control are presented. In the field of AI for robotics, the use of Large Language Models (LLMs) for enabling Human-Robot Collaboration is being discussed as a means of improving human-robot collaboration performance. AI can be introduced for monitoring and optimization of robot-assisted adhesive deposition in large parts while compensating for disturbances during the path following process. In a similar manner, semi-active exoskeletons can be made autonomous by increasing their awareness using AI-powered computer vision techniques. Deploying collaborative robots in manufacturing presents diverse challenges, thus integrating multiple risk factors into task sequencing models can improve efficacy and adaptability to diverse risk levels. Training a single model for bin picking proposes a challenge due to high-mix low-volume situations, and for that purpose a modular pipeline, which splits the problem into sub-questions, each of which refers to a separate component of the picking process, can help address this challenge. In most industrial use cases, it is difficult to collect and annotate data for small objects, as it is time-consuming and prone to human errors. Thus, synthetic data generation is discussed as a means for collection and annotation of small object datasets.

The third part is devoted to "Artificial Intelligence at Manufacturing Process Level". A manufacturing process can be defined as the use of physical mechanisms to transform the shape or properties of a material. This chapter discusses AI topics, related to manufacturing processes. A mapping is sketched between AI territories and manufacturing process-related operations by considering different aspects of the processes, such as monitoring, modeling, optimization, design and preparation, as well as control, twinning and operation. Moreover, specific cases are studied such as metal forming process with the use of AI and AI-supported vision systems for fabric defect detection and process monitoring purposes in steel factories.

We appreciate the partnership with Springer, Turnitin, EasyChair and the members of the Artificial Intelligence in Manufacturing Network (AIM-NET, www.aim-net.eu) for their essential support during the preparation of ESAIM 2023.

November 2023

Achim Wagner
Kosmas Alexopoulos
Sotiris Makris

Organization

Chairmen

Achim Wagner	German Research Center for Artificial Intelligence (DFKI), Germany
Kosmas Alexopoulos	Laboratory for Manufacturing Systems & Automation (LMS), Greece
Sotiris Makris	Laboratory for Manufacturing Systems & Automation (LMS), Greece

Organizing Committee

Katharina Hengel	German Research Center for Artificial Intelligence (DFKI), Germany
Aleksandr Sidorenko	German Research Center for Artificial Intelligence (DFKI), Germany
Zoi Arkouli	Laboratory for Manufacturing Systems & Automation (LMS), Greece

Scientific Committee

Andrea Fernández Martínez	AIMEN Technology Centre, Spain
Félix Vidal Vilariño	AIMEN Technology Centre, Spain
Ignacio Montero Castro	AIMEN Technology Centre, Spain
Santiago Grandal Montero	AIMEN Technology Centre, Spain
Santiago Muiños Landi	AIMEN Technology Centre, Spain
Caroline Vienne	French Alternative Energies and Atomic Energy Commission (CEA), France
Andrea Orlandini	Consiglio Nazionale delle Ricerche (CNR), Italy
Nicola Pedrocchi	Consiglio Nazionale delle Ricerche (CNR), Italy
Petr Kolář	Czech Technical University in Prague (CTU), Czech Republic
Pavel Burget	Czech Technical University in Prague (CTU), Czech Republic
Petr Kadera	Czech Technical University in Prague (CTU), Czech Republic

Achim Wagner	German Research Center for Artificial Intelligence (DFKI), Germany
Aleksandr Sidorenko	German Research Center for Artificial Intelligence (DFKI), Germany
Katharina Hengel	German Research Center for Artificial Intelligence (DFKI), Germany
Francois Picard	Danish Technological Institute (DTI), Denmark
Mikkel Labori Olsen	Danish Technological Institute (DTI), Denmark
Bart Meyers	Flanders Make, Belgium
Klaas Gadeyne	Flanders Make, Belgium
Danilo Brajovic	Fraunhofer IPA, Germany
Marco Huber	Fraunhofer IPA, Germany
Marco Roth	Fraunhofer IPA, Germany
Xinyang Wu	Fraunhofer IPA, Germany
Aitor Fernandez Gomez	IDEKO/Manufacturing Alliance, Spain
Gorka Unamuno	IDEKO/Manufacturing Alliance, Spain
Iñigo Bediaga	IDEKO/Manufacturing Alliance, Spain
Itziar Ricondo	IDEKO/Manufacturing Alliance, Spain
Juanan Arrieta	IDEKO/Manufacturing Alliance, Spain
Seyyed Ehsan Hashemi Petroodi	IMT Atlantique, France
Antonio Almeida	INESC TEC, Portugal
Alex Papacharalampopoulos	Laboratory for Manufacturing Systems & Automation (LMS), Greece
Kosmas Alexopoulos	Laboratory for Manufacturing Systems & Automation (LMS), Greece
Panos Stavropoulos	Laboratory for Manufacturing Systems & Automation (LMS), Greece
Sotiris Makris	Laboratory for Manufacturing Systems & Automation (LMS), Greece
Zoi Arkouli	Laboratory for Manufacturing Systems & Automation (LMS), Greece
Mostafizur Rahman	Manufacturing Technology Centre (MTC), UK
Bianca Maria Colosimo	Politecnico di Milan, Italy
Marco Luigi Giuseppe Grasso	Politecnico di Milano, Italy
Tania Cerquitelli	Politecnico di Torino, Italy
Luis Usatorre	TECNALIA, Spain
Mariluz Penalva	TECNALIA, Spain
José Luis Outón Méndez	TECNALIA, Spain
Gu van Rhijn	TNO, Netherlands
Michael van Bekkum	TNO, Netherlands
Wico Mulder	TNO, Netherlands
Wietse van Dijk	TNO, Netherlands

Heli Helaakoski	Technical Research Centre of Finland (VTT), Finland
Marko Jurmu	Technical Research Centre of Finland (VTT), Finland
Johan Stahre	Chalmers University of Technology, Sweden
Marco Taisch	Politecnico di Milan, Italy
George Michalos	Laboratory for Manufacturing Systems & Automation (LMS), Greece
Panagiotis Stavropoulos	Laboratory for Manufacturing Systems & Automation (LMS), Greece
Nikolaos Nikolakis	Laboratory for Manufacturing Systems & Automation (LMS), Greece

Contents

Artificial Intelligence at Manufacturing System Level

An Integrated Active Learning Framework for the Deployment of Machine Learning Models for Defect Detection in Manufacturing Environments

Fabián González Fragueiro, Daniel Gordo Martín, Alberto Botana López, Adrián Alonso Rial, Jacobo Otero Tranchero, Betty Cortiñas Lorenzo, Juan Manuel Fernández Montenegro, and Santiago Muiños-Landin(✉)

AIMEN Technology Centre, C/ Relva 27 A. Torneiros, 36410 Pontevedra, Porrino, Spain
santiago.muinos@aimen.es

Abstract. The digitalization of factories has triggered a humongous number of applications of Machine Learning techniques to optimize manufacturing processes. In the last years, different methodologies have been suggested for tasks like predictive maintenance, defect detection, or advanced perception systems that might contribute to a better performance of entire production lines optimizing their different steps. However, one of the main issues for the deployment of these technologies in real manufacturing environments is the management of complex tools that need, in many cases, specific AI-related knowledge from workers that are not used to work with AI-based systems. In this work we present a set of tools for defect detection tasks within a real factory based on deep learning methods and we demonstrate their performance as parts of different quality control systems. In addition, to integrate all these tools, we introduce MINT a modular intelligent framework for deep learning model management. MINT uses active learning to learn efficiently from limited labeled data, allowing manufacturers to optimize production processes with higher accuracy and reduced data annotation efforts. This framework makes more accessible to non-experts steps like model training,performance evaluation and model traceability while providing an explainability module for classification models that encourage trust and adoption of AI between factory workers. Meaning that our tool contributes to the deployment of AI techniques in real manufacturing scenarios making more accessible the management of the entire life cycle of AI models.

Keywords: Deep Learning · Quality Control · Active Learning

1 Introduction

In the current scenario of modern manufacturing, artificial intelligence (AI) has emerged as a transformative technology, revolutionizing production processes and unlocking unprecedented opportunities for efficiency, productivity, or innovation. By harnessing the power of advanced algorithms and machine learning techniques, AI has become an indispensable tool for manufacturers seeking to stay competitive in an increasingly dynamic and data-driven landscape.

A. Wagner et al. (Eds.): ESAIM 2023, LNME, pp. 3–14, 2024.
https://doi.org/10.1007/978-3-031-57496-2_1

The impact of AI on manufacturing spans a wide range of areas, from optimizing supply chains with advanced quality control methods, to enhancing predicting maintenance needs enabling intelligent automation. By analyzing vast amounts of data in real-time, AI systems can identify patterns, anomalies, and insights leading to better decision-making and resource allocation [1] [2]. While AI brings immense potential, its successful implementation and management pose a remarkable set of challenges. The management of AI models has emerged as a critical issue that manufacturers must address. As models become more sophisticated and complex, ensuring their accuracy, reliability, and ethical use becomes paramount.

One major challenge lies in the continuous training and retraining of AI models. Manufacturing environments are subject to ever-changing conditions, requiring models to adapt and evolve accordingly. Managing the vast amounts of data needed for training, as well as the computational resources and infrastructure to support the training process, demands careful planning and investment.

Another crucial aspect is the need for robust governance frameworks surrounding AI. Manufacturers must navigate ethical considerations, such as bias mitigation, privacy protection, and transparency, to ensure that AI systems operate in a fair and responsible manner. Additionally, securing AI models against potential cyber threats and ensuring compliance with data protection regulations pose additional complexities that demand diligent management.

This work introduces MINT (Modular Intelligent Network Trainer) as a solution for managing AI models in manufacturing environments, specifically focusing on its deployment in an automotive factory. MINT serves as a versatile tool for managing different AI models used in quality control across various piece types and references along the production line. The integration of active learning capabilities empowers manufacturing teams to efficiently fine-tune AI models with minimal labeled data, enabling continuous improvement and adaptability to varying production scenarios. To ensure usability for a wide range of technical knowledge, the development of MINT involved collaboration with workers from diverse backgrounds, resulting in a user-friendly tool that doesn't require expertise in AI.

The paper is structured to showcase the modularity of MINT and its practical applications. It begins by providing context, highlighting the necessity for a tool like MINT and explaining its benefits in managing AI models (active learning, deployment, monitoring and explainability). Subsequently, the paper presents several AI use cases implemented within the same factory, illustrating the diverse problems encountered and the wide array of models that MINT can handle. Each use case is introduced briefly, outlining the approach taken for image-based defect detection. Next, the paper briefly presents the main results achieved with the AI models in each use case, offering insights into their performance and how MINT facilitates their management. Finally, the paper concludes by discussing the current improvements made to the tool and outlining future perspectives for its development. The use cases covered in this work primarily revolve around binary classification tasks, specifically Tripods and Cages, as well as Tab Slots and Crack Detection in metallic parts (Fig. 1).

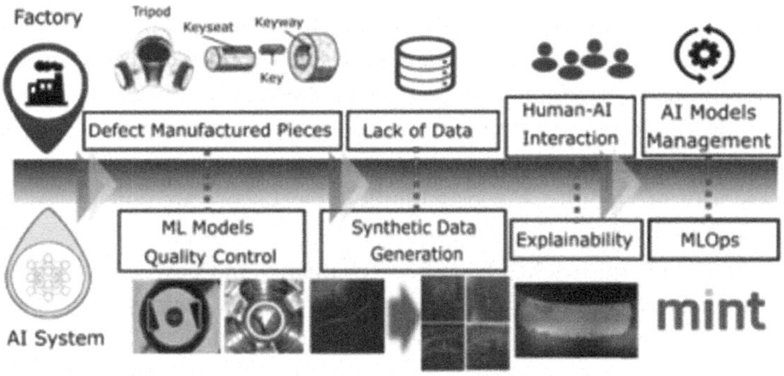

Fig. 1. Factory-AI system path. The factories within their production lines represent living labs for AI systems, generating large amounts of data. The inspection of such pieces can be addressed through different ML models. MINT allows non-experts to manage these models (training, evaluation, deployment and monitoring).

2 The Problem

Within robotic inspection cells, robotic arms, PLCs, cameras and locally running machine learning models are employed to classify pieces, ensuring quality control efficiently. Just like robots, machine learning models also require maintenance with some frequency as they may degrade over time and reduce their performance. Furthermore, the complexity of managing models in a factory escalates as their number increases. In this context, MINT is a custom application designed to ease and optimize the management of models deployed in a production plant. It provides a unified solution to various stages of the model development such as labelling, training and deployment in a centralized manner for binary and multiclass classification use cases. Additionally, MINT integrates with manufacturing cells as shown in Fig. 2, enabling communication with PLCs, and facilitating the management of data generated during automatic inspection. One of the causes of model degradation is concept drift [3]. This phenomenon refers to changes in the underlying distribution of streaming data over time (emergence of new defects, introduction of new piece references, changes in the cell environment, etc.), leading to a deviation from the original data distribution observed at the time of training. The following sections explain the modules of MINT and how they help users to manage the models deployed in multiple inspection cells of a factory.

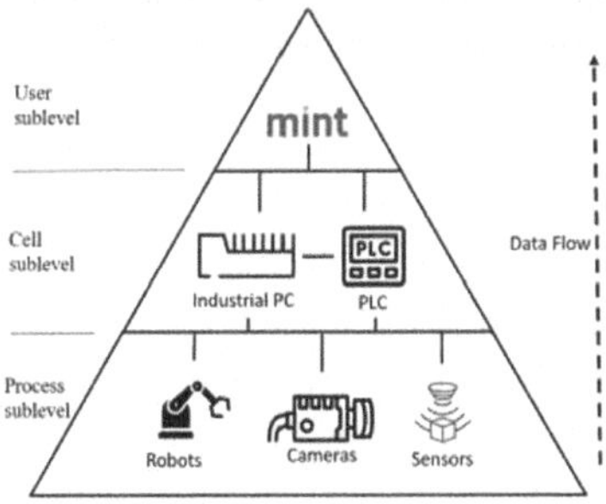

Fig. 2. Hierarchy of a robotic inspection system

2.1 Active Learning

MINT uses active learning to reduce the reliance on indiscriminate labeling [4], which is extremely resource consuming, by selectively choosing the most informative samples [5, 6]. It has three phases:

1) User feedback loop: Users are presented with batches of images which have been predicted by models in production. They are requested to provide feedback to determine, according to their expertise, whether they have been correctly classified or not. The criterion followed to present samples is based on their least confidence score [7], an uncertainty-based querying strategy that selects data whose most likely label has the lowest posterior probability. Pieces with the lowest score are considered suspicious of being misclassified.
2) Training: Once several batches have been evaluated and new samples are added to datasets, the user is given the opportunity to start a retraining cycle. Training cycles freeze the last layers of the current model and fine-tuned them.
3) Evaluation: This process involves computing the F1-score metric for the overall expanded test set to check if the new model outperforms the previous one.

2.2 Deployment

It is necessary to deploy new models in the shortest possible time, since manufacturing cells are working steadily, and production cannot be halted. MINT takes advantage of factory network to communicate with cells. When user decides to update a model, the weights file is sent through the network to the cell computer, and it starts a communication protocol with the PLC that allows real-time updating.

2.3 Monitoring

In the monitoring phase of a robotic inspection process, the calculation of metrics has two primary purposes, i.e., detecting potential deviations in the inspection process and monitoring concept drift. To identify process deviations, variables of the control process such as illumination levels and position of the inspected pieces are calculated. If any of these metrics exceed predefined thresholds, an alarm is triggered to warn manufacturing engineers to inspect the cell. Concept drift is monitored by examining the average least confidence score over a specific period. A decrease may indicate a potential model degradation.

2.4 Explainability

Explainable methods are designed to provide insights into the learning process of a neural network [8]. For classification models managed by MINT, the class activation mapping has been used [9]. In this method, the last dense output layer is removed and replaced by an average global pooling. This layer outputs a heatmap that shows which parts the network is looking at for each class.

The utilization of class activation mapping serves as a valuable tool for debugging, as it enables users to identify potential biases introduced by the model. By analyzing the heatmaps, users can gain a better understanding of how the model attributes importance to different parts of the input data, ensuring transparency and interpretability in the decision-making process.

3 Use Cases

3.1 Binary Classification

Tripods. Tripods have a key role in vehicle's movement transmission since they allow to connect two longitudinally arranged discontinuous axes. This small piece has two positions, the face and the chamfer, which are assembled by a robot. The robot picks the tripod from a conveyor belt and deposits it in a press (it must be with chamfer up) to insert it with pressure inside a joint. To date, the operator had to deposit tripods over the conveyor in a concrete way (chamfer up) to avoid wrong assemblies (those made with face up) (Fig. 3).

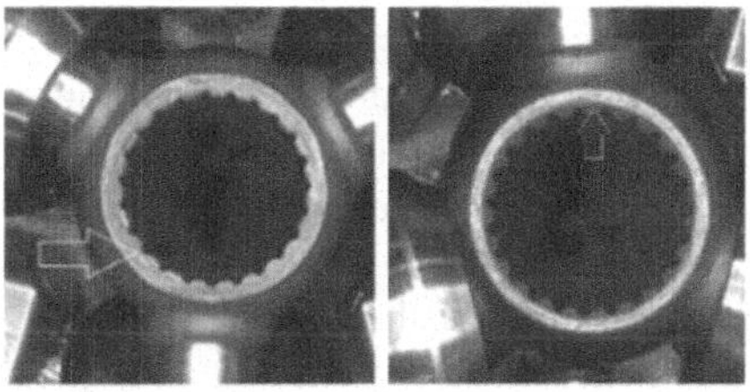

Fig. 3. (Left) Arrow pointing to tripod's face, (Right) Arrow pointing to tripod's chamfer.

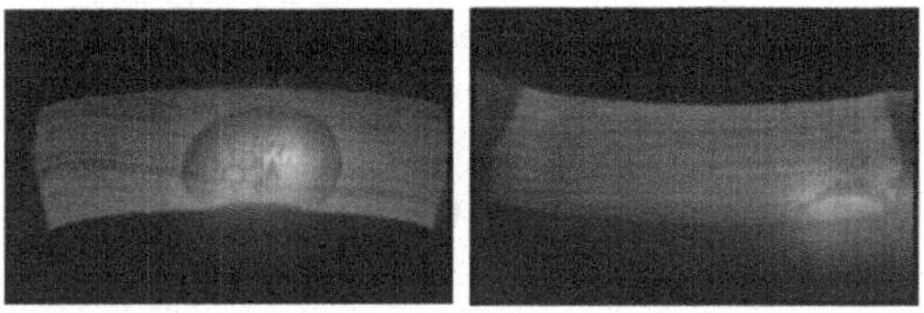

Fig. 4. Class defect heatmaps. On the left, a drop activates the neurons, while on the right it is the lack of material.

Cages. A ball cage refers to a component used to guide and retain ball bearings within the transmission system. It helps to maintain proper spacing and alignment of the balls, ensuring a smooth movement within the transmission assembly. Ball cage manufacturing involves several steps on different machines, including the punching of the ball pockets, which generate stress in the material and may lead to the appearance of defects. Defects in this piece appear on the window-like ball pockets, which will be the future interaction surface with some rolling balls. The current method of addressing this problem is visual inspection. A trained operator is responsible for checking every window manually and separate defective ones, which is a very time-consuming task. In Fig. 4 two surface heatmaps from explainability MINT module are shown.

Models. In order to address this task, a binary classification algorithm was used to detect if a piece was deposited faced up or chamfer up over the conveyor, and depending on that, the robot will pick the piece with a specific path to carry it onto the press. This involves training a convolutional neuronal network (CNN) with all used references in the assembly line. In this application a Residual Network (ResNet50) was used. Exploiting its flexibility; the possibility of modifying the architecture to fit it to a concrete application. Following this approach, one of the commonly used techniques is to freeze the learning process during the first layers. Finally, the ResNet50's output after the average pooling is modified to adapt it to the application. A fully connected layer at the end must be modified to get the number of desired class probabilities. The first 10 layer switched off during training and last fully connected layer was modified to output two classes.

3.2 Multiclass Classification

Tab Slot. Vehicle transmission is composed of several pieces which must be mounted mechanically. A very common geometry used during the assembly is the tab slot. As a high variety of references are manufactured, there are numerous different tab slot geometries, basically grouped in 4 shapes, and within each of them, there are numerous sizes and combinations. As a result, there are hundreds of different references to control.

Within defectology bibliography, defects are classified mainly into two groups: esthetic and manufacturing ones. The first group represents approximately the 70% of the rejections and includes basically dirt pieces which were not washed in the machine properly. Despite the fact that this is not of the utmost importance for the client, those pieces are rejected because the current vision system fails. The second group of rejections represents about the 30% and includes machining defects like tab slots not well geometrically defined, tab slot type error, or any kind of deviation.

Fig. 5. Existing tab slot geometries (L,E,V,U)

Commercial Vision System. To deal with this problem, a commercial vision system is installed at the end of each production line inside of an inspection cabin. Depending on the inspection veredict, a robot will pick and leave the piece into a big basket or the piece will be deviated from the main flow to a rejection zone for further supervision. This commercial vision system applies pixel based-on processing and needs a different program for each reference. As the same reference can be manufactured in multiple production lines, the default program has to be modified due to variations such as illumination or camera pose. This leads the staff to do corrections repeatedly to avoid a high impact in false rejections (a piece OK with vision verdict NOK). As a result, there are hundreds of different programs to solve the same problem for every production lines (Fig. 5).

Detection Models. To deploy a model in the inspection cabin, an industrial PC was installed, and a model was trained with 1000 images by class from all production lines. Real time open-source model YOLOv7 was used to develop this task. Ideally, the best approach is to feed the model with the most variety of labeled images for each four classes, splitting the dataset in 70% for training, 20% for validation step and 10% for testing.

After 350 epochs with early stop and batch size of 8, the reached percentage of accuracy in F1 score graphic was the 100% when confidence threshold is set up at 77%. In addition, Non Max Suppression between classes was applied to select the higher confidence detection class in case of a high Intersection Over Union value. The confidence threshold for tab slot detection was set up at 70%.

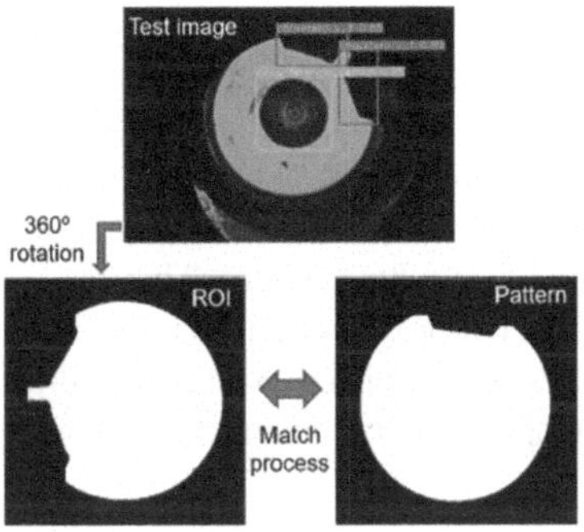

Fig. 6. Once the model is fed with a test image, inference is carried out and based on the piece's centre detection, a ROI + + is extracted and rotated 360 degrees with a pixel-wise match operation with respect to a pattern, to calculate the max match value

Due to the wide variety of references and deviations in each manufacturing line, the model must ensure that the inference is correct, and it also must match with the reference pattern in a pixel-wise comparison operation. While the photograph is taken, the piece is placed over a holder, rotated an uncontrolled degrees quantity, which means that there is an uncontrolled step which can affect detections. To achieve an extra sturdiness, several images are post-generated by detecting the piece's center during inference (Fig. 6). The center detection has two purposes; firstly, it is used to get only the piece itself (center image) and it also applies some pixel processing to avoid noise (dirt particles) from machining process. Not only is one inference done, but also a selected number of inferences by rotating the piece splitting the 360 degrees into the number of rotations selected by the operator without compromising the cycle time. Once all the inferences are done, the mode is calculated and if it is higher from a configurable percentage and the result matches with the ground truth class, the result is OK. Conversely, the piece is rejected by sending the result to the PLC.

3.3 Object Detection

Piece and Defect Morphology. On one hand, pieces present a cup-shaped morphology in which target defects only appear inside the body of the cup, which forces the system to take several images from different angles to examine the whole surface. Each piece belongs to a different part family that share similar size and geometry. On the other hand, the defects appear in the form of cracks. A crack can be defined as a local discontinuity of material and it appears as a small, elongated, often straight line on the surface of the piece. There were 6 different types of cracks as shown in Fig. 7. These cracks clearly reduce the strength of the material and are the cause of fatigue failure during the product lifespan [10]. Therefore, it is vital to detect them before they continue to the next phase of the production process.

Non-destructive Testing: Magnetic Particle Inspection. The current way of addressing the problem is by a non-destructive testing method called Magnetic Particle Inspection [11]. In this method, the piece is exposed to a magnetic field and then coated with a ferrous particle solution that accumulates in the surface whenever it meets a discontinuity. The main problem is the strong dependence on the qualifications and concentration

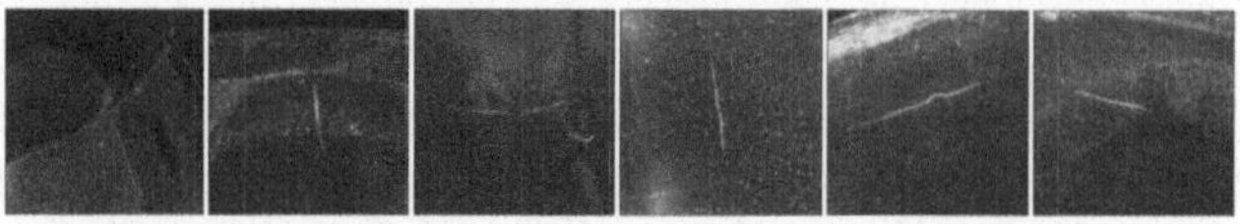

Fig. 7. Types of defects.

of the inspector [12], which decreases during the working day because of visual fatigue. A blind [16] experiment was accomplished to assess the performance. This experiment involved a group of 4 inspectors with varying degrees of experience who were tasked to inspect the same set of components at the beginning and at the end of their working day. This set was composed of 10 defective pieces and 50 without defect. The results revealed a significant disparity in performance between the more experienced inspectors and their less experienced counterparts. Notably, the less experienced exhibited poorer performance, particularly towards the end of their workday when visual fatigue was more pronounced.

Detection Models. Throughout this study two distinct models have been used and compared: a FCNN [13] and a YOLOv7 [14] model. The FCNN model, built upon ResNet50, was trained patchwise, involving the extraction of 224×224 patches containing defects from the dataset, along with background samples of the same dimensions. To determine whether a sub-image represents a crack or non-crack, a final layer consisting of a single neuron was utilized to output the probability. A generative adversarial network was added to generate realistic samples as a form of data augmentation [15]. On the other hand, YoloV7 provides the advantage that its training is fully convolutional, so full images are shown to the network during training. Thus, the model has more contextual information about the surroundings of the defect which improves detection. It was trained during 20 epochs with the basic data augmentation.

4 Results

4.1 Binary Classification

4.2 Multiclass Classification

In order to get an overall vision of both systems and their performances, a test was evaluated over a month production. Thus, as it is shown in Fig. 8 below, 42374 pieces belonging to eight different references but only two geometries (v,u) were manufactured and inspected by both systems at the end of line production with only a 0,3% of false positives for YoloV7 vs 0,7% gotten by the commercial vision system. It is worth noting that the 75% of these 42374 pieces belong to v-geometry and the rest to u-geometry. Regarding the precision, Fig. 8 depicts the performance achieved by the application developed by AIMEN vs the commercial vision system, which in case of AIMEN, was higher in all manufactured references. The main difference lies in YoloV7 achieves these results with the same program instead of having numerous different programs for the same task.

Table. 1. The accuracy on the test set is about 100% in all references, except that with less samples in the train dataset (ref. C), in which the accuracy dropped slightly.

REF	Orientation	Image Quantity	Accuracy (%)
A	Face	1324	100
	Chamfer	2093	100
B	Face	1680	99
	Chamfer	2035	100
C	Face	25	88,23
	Chamfer	1235	100
D	Face	179	100
	Chamfer	1141	100

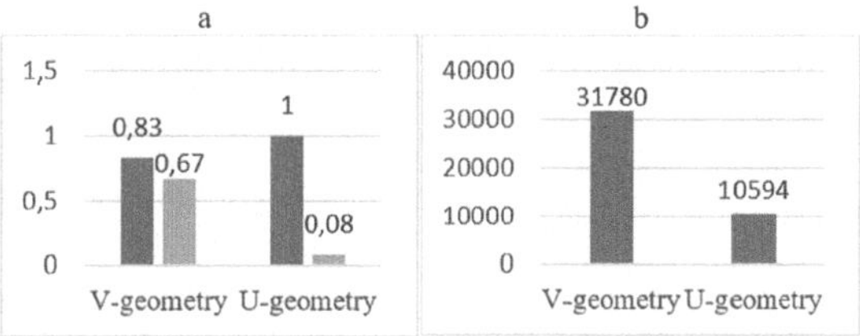

Fig. 8. **a)** Precision achieved by the application developed by AIMEN (blue) vs commercial vision system (green) in two manufactured references. **b)** Dataset quantity for each geometry.

4.3 Object Detection

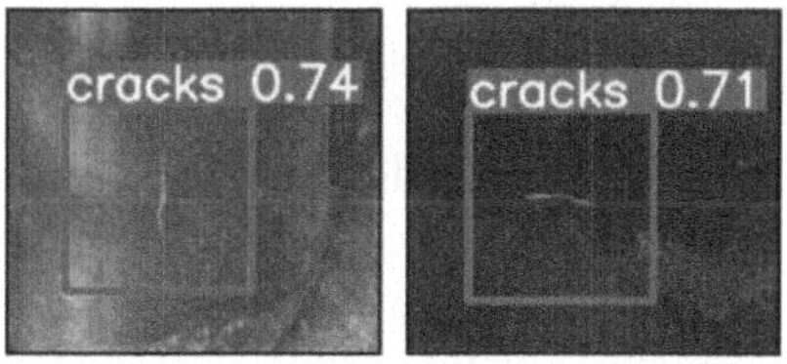

Fig. 9. Detection of cracks in the test set.

The f1-score metric was employed to compare the performance of the two models. Both models were tested on the same validation set, comprising diverse part families and crack typologies. Results indicated that the yolov7 model exhibited superior performance, achieving a f1-score of 0.93 versus a 0.89 for the FCNN. In practical production scenarios, this metric translates to an almost perfect correct classification of all defective pieces, considering that each defect is present in multiple positions as previously mentioned. Therefore, the compound probability of detecting a crack in a defective piece tends to approach 1, ensuring high accuracy. In Fig. 9 some examples of detections for test set images are shown.

4.4 MLOps

The modularity and effectiveness of MINT has been addressed in two of the use cases (binary classification) while further improvement is in progress to integrate labeling functionalities using CVAT for multiclass problems. Using MINT, once the training is completed and a new model demonstrates improved performance compared to the previous one, it is stored within the application's model manager. Utilizing this tool, users can select a specific option from the menus when booting up a use-case. This selection prompts a dialogue box displaying a drop-down menu that lists all previously recorded models, along with their corresponding F1-Score metric.

The dialogue box provides users with the option to select and load a previous model as the active model. This functionality is particularly valuable in scenarios where additional testing needs to be conducted on a previous model or when there is a desire to explore a different direction based on a previous checkpoint, rather than beginning a new. By allowing users to load past models as the active model, the tool promotes flexibility and efficiency in the development and refinement of AI models. It eliminates the need to start from scratch, enabling users to build upon prior work and leverage the knowledge gained from previous iterations. This capability empowers users to conduct in-depth analysis, fine-tuning, and exploration while benefiting from the progress made with

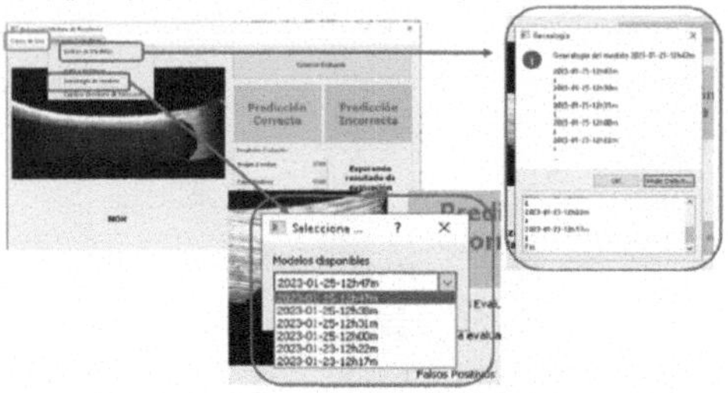

Fig. 10. MINT main view. Different use cases (green box) can be loaded. MINT can import any model (blue box) associated with the use cases whose traceability is guaranteed by the genealogy management (red box).

In addition to the model manager window, the application also includes a complementary genealogy viewing tool. When the user accesses this function, a similar drop-down menu is presented. However, upon selecting a model, a second window appears, displaying a chronological list of dates and metrics for the preceding models. This genealogy viewing tool allows users to trace the lineage of models, starting from the current one and moving backward in time. The list presents the dates and metrics of the model immediately preceding the active one, followed by its predecessor, and so on, until reaching the first model created from a blank slate, without any predecessor.

This feature proves useful when users need to revert to a previous model but are uncertain about the specific order and relationships between different iterations. By providing this additional information, the genealogy viewing tool offers clarity and aids users in making informed decisions regarding their model selection. It allows them to navigate the model history and understand the progression of changes, ensuring confidence in choosing the desired previous model for further experimentation or exploration (Fig. 10).

5 Conclusions

In this work we presented the development of a modular application designed for the management of AI models. The application has been successfully demonstrated through various use cases, showcasing its versatility and effectiveness. The associated AI development for each use case has been thoroughly explained, highlighting the strategies employed for achieving image-based defect detection.

The results obtained from the AI models within each use case demonstrate promising performance, indicating the efficacy of the developed application. The modular approach allows for easy integration and management of different AI models, providing flexibility and adaptability to manufacturing environments. Overall, this work contributes to the field of AI model management in manufacturing by providing a user-friendly tool that can handle diverse piece typologies and references along the production line. The demonstrated performance and successful implementation of the application underscore its potential for enhancing quality control processes in the automotive sector and other manufacturing industries.

6 Aknowlegments

This work has been developed within the frame of the Joint Research Unit "F4ctorIA" between GKN Driveline Vigo and AIMEN Technology Centre, granted by the Galician Innovation Agency and co-funded by the European Union through the European Regional Development Fund (ERDF) with reference IN853A-2020/02. Operational Program FEDER GALICIA 2014–2020.

References

1. Stavropoulos, P., Papacharalampopoulos, A., Sabatakakis, K.: Robust and Secure Quality Monitoring for Welding through Platform-as-a-Service: A Resistance and Submerged Arc Welding Study. Machines **11**(2), 298 (2023)
2. Xia, C., Pan, Z., Polden, J.L.H., Xu, Y., Chen, S., Zhang, Y.: A review on wire arc additive manufacturing: Monitoring, control and a framework of automated system. J. Manufact. Syst. **57**, 31–45 (2020)
3. Bayram, F., Ahmed, B.S., Kassler, A.: From Concept Drift to Model Degradation: An Overview on Performance-Aware Drift Detectors (2022)
4. Hino, H.: Active Learning: Problem Settings and Recent Developments. arXiv:2012.042 25v2, (2020)
5. Aggarwal, C.C., Kong, X., Gu, Q., Han, J., Philip, S.Y.: Active learning: a survey. Data Classification, p. 599–634 (2014)
6. Zhan, X., Wang, Q., Huang, K.-h., Xiong, H., Dou, D., Chan, A.B.: A Comparative Survey of Deep Active Learning (2022)
7. Lu, J., Liu, A., Dong, F., Gu, F., Gama, J., Zhang, G.: Learning under concept drift: a review. IEEE Trans. Knowl. Data Eng. **31**(12), 2346–2363 (2020)
8. Gohel, P., Singh, P., Mohanty, M.: Explainable AI: current status and future directions, Arxiv (2021)
9. Zhou, B., Khosla, A., Lapedriza, A., Oliva, A., Torralba, A.: Learning Deep Features for Discriminative Localization (2015)

10. Zerbst, C.K.U.: Material defects as cause for the fatigue failure of metallic components. Inter. J. Fatigue, 312–323 (2019)
11. Mouritz, A.P.: Nondestructive inspection and structural health monitoring of aerospace materials. Introduction to Aerospace Materials, pp. 534–557. Woodhead Publishing
12. Goebbels, K.: A new concept of magnetic particle inspection. In Non-Destructive Testing, Amsterdam (1989)
13. Jonathan Long, E.S.T.D.: Fully Convolutional Networks for Semantic Segmentation (2014)
14. Wang, C.-Y., Bochkovskiy, A., Liao, H.-Y. M.: YOLOv7: Trainable bag-of-freebies sets new state-of-the-art for real-time object detectors, Arxiv (2022)
15. Lopez, A.B.: GAN-based data augmentation (2021). https://doi.org/10.5281/zenodo.7074639
16. Enkvist, J., Edland, A., Svenson, O.: Operator Performance in a Blind Test Piece Trial, vol. 59, pp. 531–536
17. Simon, G., Vakulya, G., Rátosi, M.: The way to modern shutter speed measurement methods: a historical overview. Sensors (Basel), **22**(5) (2022)
18. Saberironaghi, A., Ren, J., El-Gindy, M.: Defect detection methods for industrial products using deep learning techniques: a review. Algorithms **16**(2), 16, 95 (2023)

Complex and Big Data Handling and Monitoring Through Machine Learning Towards Digital-Twin in High Precision Manufacturing

Marco Grasso[1](✉), Giuseppe Fogliazza[2], and Bianca Maria Colosimo[1]

[1] Department of Mechanical Engineering, Politecnico di Milano, Via La Masa 1, 20156 Milan, Italy
marcoluigi.grasso@polimi.it

[2] MCE Solutions, Via F. e G. Celaschi, 19, 29020 Vigolzone, PC, Italy

Abstract. The digital transition faced in the Industry 4.0 framework is reshaping the complexity and volume of data used for discrete manufacturing quality monitoring and modelling. Advanced manufacturing methods are used to produce more and more complex products, while the wide adoption of in-line sensing tools enable big and fast streams of sensor data while the part is being produced. The result is a novel level of quality data complexity, which gives rise to industrial challenges, namely the lack of appropriate statistical process monitoring methods, the difficulty to manage such high dimensionality of in-process and post-process data, and the difficulty to deal with multi-sensor data stream in an efficient and effective way. This paper presents novel solutions based on machine learning (ML) algorithms suitable to fuse and make sense of big multi-channel data gathered in discrete manufacturing applications to anticipate the detection of anomalies, and to enhance the quality of the products, towards the implementation of digital-twin solution in production environments. A real case study is presented in the field of milling operations via flexible systems. ML methods are compared against benchmark statistical techniques to highlight when and to what extent the actual ML potentials may help bridge the gap between scientific research and industrial deployment, pointing out current barriers and challenges in real production environments.

Keywords: Industry 4.0 · multichannel signals · Recurrent Neural Network · process monitoring · digital twin

1 Introduction

The Industry 4.0 paradigm has reshaped the production environments, opening a variety of new opportunities and bringing novel tools and manufacturing methods into more and more digitalized factories. Grounding on this, the discrete manufacturing industry is currently facing a new evolution, also known as green and digital (twin) transition. The increasing flexibility and digitalization of manufacturing systems give rise to a new availability of big and complex quality data in every stage of the production. In this

A. Wagner et al. (Eds.): ESAIM 2023, LNME, pp. 15–23, 2024.
https://doi.org/10.1007/978-3-031-57496-2_2

framework, companies must tackle the inflated dimensionality, veracity and complexity of quality data needed to monitor both the process stability and the health conditions of manufacturing systems over time [1]. Digital factories are more and more equipped with sensors embedded into production systems, suitable to provide high-speed and high-resolution data in several different formats for the whole duration of the manufacturing process. These data may be profitably used while the part is being produced, to anticipate the detection of anomalies, defects, and unstable process states, enabling novel in-process and in-line monitoring capabilities.

The increased quality data complexity (high-dimensional multi-stream real-time signals, heterogenous multi-source signals, high-speed videos, lack of repeatable signal patterns, etc.) affects all production methods, from traditional discrete manufacturing processes to additive manufacturing solutions, giving rise to several new challenges [2, 3]. However, there is a lack of efficient and robust quality modelling and monitoring methods suitable to handle high-dimensional data that can be gathered in-line, e.g., in the form of multi-stream and time-varying signals characterized by complex patterns [4, 5]. This study explores how machine learning and artificial intelligence (AI) techniques may be profitably used in this framework, to monitor multi-stream signals gathered either during the cutting process or in no-load machine tool operations. Presented examples include low-dimensional learning methods applied to multi-channel profile data, and multi-stream data pattern prediction via recurred neural networks (RNNs), i.e., the Gated-Recurrent-Unit. The possibility to move from purely data-driven to digital twin solutions is finally discussed in the last section, opening to further extensions of proposed solutions. Presented solutions have been patented (or patent-pending) in collaboration with MCM (Maching Centers Manufacturing) SpA, and they have been validated in real production. Therefore, the present study aims to provide examples and case studies that are representative of how the AI gap between scientific research and industrial adoption can be bridged, outlining also the open issues and direction for future developments and improvements. Section 2 presents a quick overview of the state-of-the-art. Section 3 introduces the case studies. Section 4 and 5 presents the proposed methodologies and salient results. Section 6 conclues the paper.

2 Brief Overview of the State-of-the-Art

A vast literature has been devoted to in-line process monitoring in discrete manufacturing, with first seminal papers that dates back to the 70's [6–8]. Nevertheless, there is still a broad gap between advanced methodologies explored in the literature and nowadays available industrial solutions. Almost all industrial toolkits for process monitoring from major third-party developers, namely Marposs, Artis, Komet Group, MCU, Sequoia, Montronix, Sandvik/Prometec, Nordmann, relying on conservative data processing and heuristic alarm rules, without real data fusion capabilities. Such alarm rules are highly dependent on the human operator's experience. The aim to minimize or avoid false alarms, commonly impose conservative choices and represents a barrier against the adoption of more innovative solutions [9]. Off-the-shelf process monitoring toolkits are mostly suitable for very repeated and simple process operations (like tapping, drilling and repeated face or contour milling). Real industrial processes, instead, are strongly time-varying, characterized by non-stationary and complex signal patterns, affected by several nuisance factors, with a high natural variability, etc. [6–9]. These conditions make industrial monitoring toolkits hardly applicable in a wide range of real industrial processes [9]. As far as the scientific literature on tool, process and machine health condition monitoring is concerned, several different families of methods have been studied and tested, largely relying on AI [6–9]. However, a broad gap still exists between the literature in this field and real industrial applications. Among the most relevant reasons we can mentioned the following: 1) the large majority of studies used sensing methods (e.g., cutting forces) that are hardly or impossible to have in shop floors [9], 2) the majority of studies where tested under controlled conditions in lab environments [6–9], 3) they involved simple test specimens and simple tool trajectories [6–10], 4) they assume sufficiently large training datasets could be available [9, 10]. Additional barriers include [11]: low system reliability, limited accuracy of tool wear estimation, especially in the presence of complex processes / products, and high false alarms caused by several nuisance factors. Moreover, most AI methods studied in the literature require extended training phases and/or frequent recalibration procedures that make them poorly attractive for an actual industrial implementation. The solutions presented in this study were motivated by the aim to tackle both process monitoring and machine health condition monitoring problems in real production, with methods that are suitable for direct industrial adoption.

3 Real Case Studies

Two real case studies have been explored. The first was implemented on a Clock 700 machine tool developed by MCM, and it involved the performance of so-called "fingerprint cycles", i.e., the periodic execution of a repeated sequence of movements involving all machine exes in no-load conditions [12, 13]. Each fingerprint cycle consisted of five sequential operations: i) rapid movements in both directions along X, Y and Z direction, ii) rapid rotations of the rotary axes B and iii) spindle rotations at different speeds. The signals acquired during the execution of fingerprint cycles include the spindle speed, the spindle power absorption, and the drive motor currents of the X, Y, Z and B axes. All signals were acquired from embedded sensors via the programmable logic controller (PLC) of the system. The acquisition frequency was 20 Hz for all the signals. Figure 1 (left panel) shows an example of fingerprint signals.

The second case study regards the process monitoring of milling operations, where in-process signal data were used to keep under control the stability of the cut and to possibly detect out-of-control deviations. Also in this case, multi-stream signals from embedded sensors were used, but they were acquired during material cutting. The manufacturing process involved the series production of vacuum pumps. The process involved a contour milling along a complex trajectory with time-varying radial and axial depth of cut. An example of acquired signal patterns is shown in Fig. 1 (right panel). They were acquired through the PCL of the system with an acquisition frequency of 20 Hz. The process was carried out by means of a MCM Clock Auto machine tool. Copies of the same part were milled by using the same tool copy. The complete execution of the contour milling on one part was denoted as process cycle.

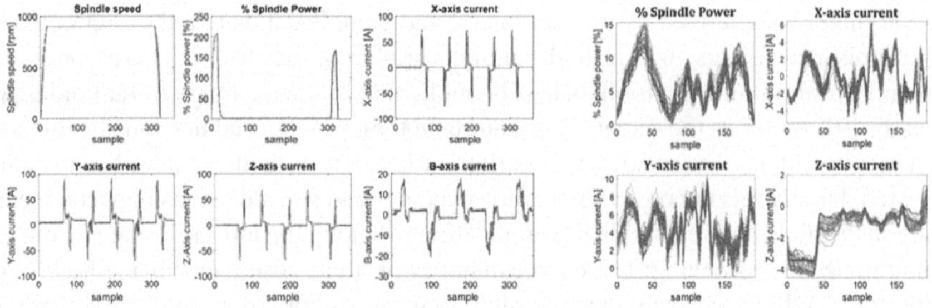

Fig. 1. Left: examples of finerprint signals acquired through embedded machine sensors; right: superimposition of signal patterns acquired in consecutive milling cycles using the same tool copies (red lines refers to early tool life, blue line to approaching end-of-life).

4 Proposed Methods

4.1 Low-Dimensional Learning for Machine Health Condition Monitoring

The first proposed approach has been designed and petented for continuous, robust and automated machine health condition monitoring in flexible manufacturing systems [14]. The underlying idea consists of the so-called "fingerprint approach", i.e., using signal patterns acquired in periodically repeated operations in no-load conditions to estimate the current health state of the machine tool and its components [12, 13, 15].

The aim of the method consists of detecting degraded states of machine tool components since their onset stage, to enable an efficient implementation of maintenance operations. Three salient aspects characterize the proposed solution: i) the isolation of the production process through the implementation of periodic fingerprint cycles where all the transmission axes and the spindle are activated sequentially, ii) the acquisition of sensor signals from embedded sensors, already available from the PLC of the machine tool, and iii) the application of a low-dimensional learning methodology, i.e., Principal Component Analysis (PCA), to quickly detect changes in the multi-channel fingerpring signal pattern. Successively, the degraded performances of the machine tool axes and

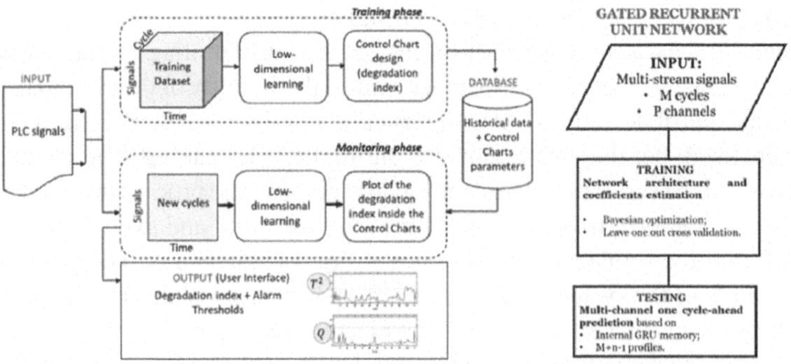

Fig. 2. Left: Scheme of the proposed low-dimensional learning and machine health monitoring approach; Right: Scheme of the proposed GRU-based approach for enhanced process predict

spindle are monitored via a PCA-based multivariate control chart. The major steps of the proposed approach are summarized in Fig. 2 (left panel). The PCA was applied on a set of representative synthetic descriptors computed for each acquired signal (i.e., time-series moments as well as peak ampliture computed in sub-portions of each cyles).

4.2 Recurrent NNs for Multi-stream Process Pattern Prediction

The second proposed approach aims at enhancing the prediction of the signal patterns in the next production cycle (i.e., during the production of the next component), based on current and past acquired signals. By predicting the signal pattern, one could detect any unexpected change caused by out-of-control conditions, like tool chipping. The challenge tackled by this research lies in applying the RNN framework – commonly designed and applied to univariate time series forecasting problems – to multi-stream signal data, opening to a generalization of the method in a wider range of industrial applications. A scheme of the method is shown in Fig. 2 (right panel). The study compares the prediction capabilities of two modeling approaches: i) the proposed approach, namely the GRU Network [16], ii) a parametric modeling solution acting as a competitor, i.e., Discrete Wavelet Transform (DWT) [17]. The wavelet modelling competitor was selected as one of most effective methods to model complex profile data patterns. GRU uses a simple structure and it features an efficient training process [16]. It requires few training data points and less computational burden than othe RNNs [18]. The training phase is aimed at finding the optimal GRU network configuration both in terms of hyperparameters and coefficients. Bayesian optimization was proposed to carry out this task, as it overperforms simpler optimization algorithms and, contrarily to the more elaborated metaheuristic methods, it is suitable for dealing with small configuration spaces [19].

5 Results

5.1 Low-Dimensional Learning for Machine Health Condition Monitoring

For each linear and rotating axes, 30 synthetic descriptors were computed in each fingerprint cycle. The PCA was applied on these descriptors for each axis and the spindle. The explained variability was set to 80%, leading to retaining up to six principal components. Figure 3 (left panel) shows an example of automated detection of a real degradation state of health of the B-axis caused by a collision during milling operations. The training phase was composed by 120 cycles performed by the end-user after a maintenance intervention. During the testing phase, a false alarm rate of 0.4% was observed over more than 500 monitored fingerpring cycles, whereas 4 real degraded states of linear and rotary axes were correctly detected, including the one in Fig. 3. The use of technology-driven synthetic descriptors also allowed a easy identification of root causes in all cases.

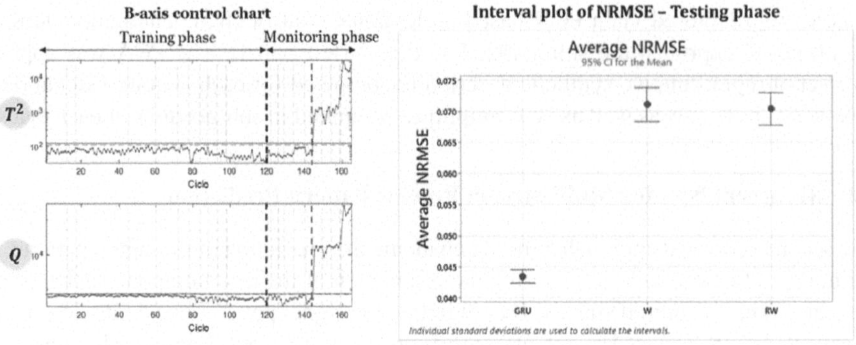

Fig. 3. Left: Example of automated detection of a real degraded state of one machine axis via low-dimensional learning; Right: performance comparison between the GRU and the wavelet-based prediction performances in the non-stationary milling process

5.2 Recurrent NNs for Multi-stream Process Pattern Prediction

The proposed GRU-based approach was tested in both real case studies described in Sect. 2. In the former case, the GRU was trained on 60 fingerprint cycles and then used to predict the multi-stream signal patterns in remaining cycles. In the latter case, the GRU was trained on 46 milling cycles, i.e., during the production of 46 copies of the same part. The goodness of the one-cycle-ahead prediction was determined by estimating the normalized root mean squared prediction error (NRMSE). The comptetitor consisted of a multi-channel adaptation of the wavelet modelling technique applied to repeating waveform signals. A Daubechie mother wavelet of order 11 was used, and implemented in two versions: in the first case, the wavelet model was fitted with training data and then used to draw predictions in following cycles; in the second case, the wavelet model was recursively updated as new cycles became available. The wavelet-based prediction was more accurate in the first real case study, where signal data exhibited a stationary pattern over time, whereas the GRU outperformed the wavelet-based approach in the second case study, because of the non-stationary behavior of the milling process (Fig. 3). This highlights that GRU can be profitably adopted in time-varying and non-stationary conditions, leading to better predictions and hence to better process monitoring performances. Nevertheless, parametric techniques are still more effective if complex data patterns regularly repeats over time.

6 Discussion and Conclusion

The former two examples involved repeated operations, i.e., periodic fingerprint cycles or series production of copies of the same part. Real manufacturing applications include much more challenging applications where the tool path as well as the instantaneous chip thickness and the material removal rate continuously change during the machining operation. This situation is typical of many industrial products (as an example, see the impeller shown in Fig. 4), where each copy of the tool cut complex geometrical features, with no repeating pattern, neither during the life of each single copy of the tool nor from one copy to another. In similar cases, the naturally varying "signature" of the process may completely mask any actual tool degradation as well as other undesired anomalies. No industrial toolkit is suitable to monitor the process under these premises, and there is an almost complete lack of methods even in the scientific literature [8–11, 20, 21]. One way to tackle this problem consists of combining real process data with process simulations, in a digital-twin oriented approach. This can be achieved by estimating a tool degradation index combining real process signals (acquired from embedded sensors during the cut) and simulated process quantities (namely, time-varying axial and radial cutting depths, chip thickness, material removal rates,...) [22]. This allows extending process monitoring capabilities to any type of machining operation, regardless of the complexity of the product and the tool path trajectory. A seminal idea is shown in Fig. 4, where the combination of real data with the output of the process simulation in a virtual environment allowed estimating the tool conditions during the milling of aluminium and steel impellers [22]. This approach can be regarded as an extension of the methods presented in the previous sections. In the Industry 4.0 framework, the increasing flexibility and capabilities of manufacturing systems gives rise to novel production paradigms that enable the creation of innovative types of products as well as the collection of new and complex quality data. Despite the wide literature devoted to process monitoring and AI solutions, there is still a broad gap between research and actual industrial implementations. The methods presented here were tested in real production environments, in collaboration with MCM SpA. Proposed methods showed to be effective in deal with real multi-channel complex signal patterns, but even higher data and process complexity can not be tackled with data-driven method only. In a continuously evolving digital manufacturing framework, digital-twin and hybrid modelling techniques represent the key to move towards a new generation of intelligent high-precision manufacturing systems. Seminal studies in this direction may open to new best practices and demonstrators with a high potential to increase the European industry competitiveness.

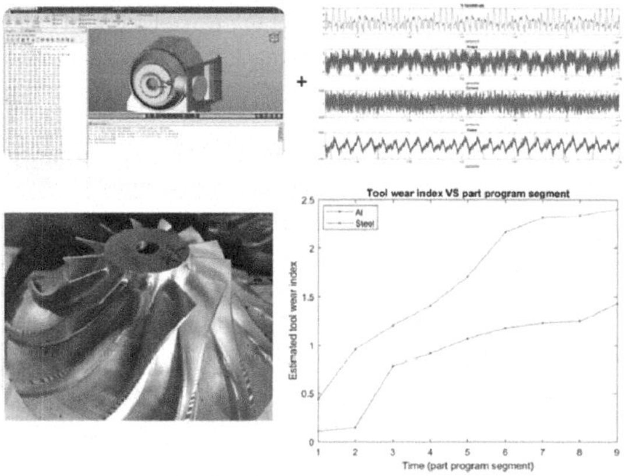

Fig. 4. Example of combination of real and simulated data for the in-process estimation of the tool conditions during the machining of an impeller (estimated tool wear indexes are shown for aluminium and steel impeller test cases)

Additional challenges shall be faced to enhance the industrial breakthrough of these and similar methods, from the implementation of adaptive and transfer learning to studying novel ways to embed physics and engineering knowledge in AI models, also moving from black-box to gray or white-box paradigms. These challenges will guide the future research in this field.

Acknowledgements. The authors acknowledge MCM SpA for kindly providing the real case studies and for the research collaboration that led to the results presented in this study. The authors also acknowledge Federica Garghetti, as the study here presented is part of her PhD thesis. Finally, authors are grateful to Prof. Massimo Pacella for his contribution to the study on the GRU methodology and its application.

References

1. Bécue, A., Praça, I., Gama, J.: Artificial intelligence, cyber-threats and Industry 4.0: challenges and opportunities. Artif. Intell. Rev. **54**(5), 3849–3886 (2021)
2. Reis, M.S., Gins, G.: Industrial process monitoring in the big data/industry 4.0 era: from detection, to diagnosis, to prognosis. Processes **5**(3), 35 (2017)
3. Papadopoulos, T., Singh, S.P., Spanaki, K., Gunasekaran, A., Dubey, R.: Towards the next generation of manufacturing: implications of big data and digitalization in the context of industry 4.0. Prod. Plann. Control **33**(2–3), 101–104 (2022)
4. Peres, R.S., Jia, X., Lee, J., Sun, K., Colombo, A.W., Barata, J.: Industrial artificial intelligence in industry 4.0-systematic review, challenges and outlook. IEEE Access **8**, 220121–220139 (2020)
5. Zhong, R.Y., Xu, X., Klotz, E., Newman, S.T.: Intelligent manufacturing in the context of industry 4.0: a review. Engineering **3**(5), 616–630 (2017)
6. Li, X., Liu, X., Yue, C., Liang, S.Y., Wang, L.: Systematic review on tool breakage monitoring techniques in machining operations. Int. J. Mach. Tools Manuf **176**, 103882 (2022)
7. Zhou, Y., Xue, W.: Review of tool condition monitoring methods in milling processes. Int. J. Adv. Manuf. Technol. **96**(5), 2509–2523 (2018)
8. Serin, G., Sener, B., Ozbayoglu, A.M., Unver, H.O.: Review of tool condition monitoring in machining and opportunities for deep learning. Int. J. Adv. Manuf. Technol. **109**, 1–22 (2020)
9. Zhou, Y., Xue, W.: Review of tool condition monitoring methods in milling processes. Int. J. Adv. Manufact. Technol. **96**, 2509–2523 (2018)
10. Mohamed, A., Hassan, M., M'Saoubi, R., Attia, H.: Tool condition monitoring for high-performance machining systems—a review. Sensors **22**(6), 2206 (2022)
11. Bombiński, S., Kossakowska, J., Nejman, M., Haber, R.E., Castaño, F., Fularski, R.: Needs, requirements and a concept of a tool condition monitoring system for the aerospace industry. Sensors **21**(15), 5086 (2021)
12. Fogliazza, G., et al.: Fingerprint analysis for machine tool health condition monitoring. IFAC-PapersOnLine **54**(1), 1212–1217 (2021)
13. Ferreiro, S., Konde, E., Fernández, S., Prado, A.: Industry 4.0: predictive intelligent maintenance for production equipment. In: European Conference of the Prognostics and Health Management Society, no (1–8) (2016)
14. Colosimo, B.M., Grasso, M.L.G., Fogliazza, G.: U.S. Patent Application No. 17/440,012 (2022)
15. Moore, J., Stammers, J., Dominguez-Caballero, J.: The application of machine learning to sensor signals for machine tool and process health assessment. Proc. Inst. Mech. Eng. Part B J. Eng. Manuf. 0954405420960892 (2020)

16. Dudek, G., Smyl, S., Pełka, P.: Recurrent Neural Networks for Forecasting Time Series with Multiple Seasonality: A Comparative Study. arXiv preprint arXiv:2203.09170(2022)
17. Woodall, W.H.: Current research on profile monitoring. Production **17**, 420–425 (2007)
18. Qashqai, P., Zgheib, R., Al-Haddad, K.: GRU and LSTM comparison for black-box modeling of power electronic converters. In: IECON 2021–47th Annual Conference of the IEEE Industrial Electronics Society, pp. 1–5. IEEE (2021)
19. Yang, L., Shami, A.: On hyperparameter optimization of machine learning algorithms: theory and practice. Neurocomputing **415**, 295–316 (2020)
20. Iliyas Ahmad, M., Yusof, Y., Daud, M.E., Latiff, K., Abdul Kadir, A.Z., Saif, Y.: Machine monitoring system: a decade in review. Int. J. Adv. Manufact. Technol. **108**, 3645–3659 (2020)
21. Mohanraj, T., Shankar, S., Rajasekar, R., Sakthivel, N.R., Pramanik, A.: Tool condition monitoring techniques in milling process—a review. J. Market. Res. **9**(1), 1032–1042 (2020)
22. Grasso, M., Garghetti, F., Colosimo, B.M., Fogliazza, G., Ratti, A.: Metodo per stimare uno stato di un'utensile durante una lavorazione per asportazione di truciolo, National patent submission n. 102023000013071 (2023)

On Bringing Fault Detection to Skill-Based Production

Pascal Rübel[1(✉)], William Motsch[2], Hendrik Schäfer[2], and Martin Ruskowski[1]

[1] German Research Center for Artificial Intelligence (DFKI), Kaiserslautern, Germany
pascal.ruebel@dfki.de

[2] Technologie-Initiative SmartFactory KL e.V., Kaiserslautern, Germany

Abstract. In the context of Industrie 4.0, modular and skill-based production enables the manufacturing of small batch sizes. This leads to increased complexity of dynamic manufacturing contexts due to more constellations of cyber-physical production modules, skills, and products. However, even existing approaches aiming to enhance efficient and resilient production, fault detection applications, usually requiring big amounts of data or knowledge, remains mainly unnoticed. Against this background, it is the objective of this paper to develop a model that extends skill-based production by means of incorporating automated fault detection. This includes the identification of required fault detection components as well as its integration into the process control for executing skills.

Hence, to achieve this objective, state-of-the-art approaches of fault detection are adapted and integrated into the skill model. Finally, the elaborated model is prototypically implemented in a realistic demonstration environment to come up with a proof of concept. Since it is difficult to directly model faulty behavior of Cyber-Physical Production Modules in flexible production their nominal behavior during the execution of skills is modelled and compared with its actual behavior. Based on deviations and defined thresholds symptoms are generated. The developed components are synchronized with their skill execution.

Keywords: skill-based production · fault detection · flexible manufacturing · modular production

1 Introduction

Industrie 4.0 and its concepts tackle the challenge of high demand on flexibility and individualization of products and thus small batch sizes [1, 2]. Modular and skill-based production are key concepts to enable the manufacturing of small batch sizes in technical and economical manners, despite increasing complexity in terms of different constellations of production modules, skills and products [1, 3].

A lack of appropriate quality and quantity of data as well as the constantly changing composition of production entities represent challenges to resilient and efficient production. Although different approaches from the field of skill-based production address these challenges, fault detection remains unseen.

A. Wagner et al. (Eds.): ESAIM 2023, LNME, pp. 24–34, 2024.
https://doi.org/10.1007/978-3-031-57496-2_3

Taking this into consideration the objective of this paper is to develop a model that extends skill-based production by means of incorporating automated fault detection. This includes the definition and development of components required for the fault detection, extension of the skill-model as well as its integration into the process control of executed skills.

Fig. 1. Fault Detection and Diagnosis Pipeline

Conventional fault detection applications can be split up into data-driven, model- or knowledge-based methods [4]. These methods cover different stages of the fault detection and diagnosis pipeline. This work only considers the tasks related to the fault detection step (see Fig. 1). Fault detection in the context of this work is defined as the determination of faults present in a system and the time of detection. A fault is defined as an unpermitted deviation of at least one characteristic property or parameter of the system from an acceptable condition [5].

Usually, those methods depend on large amounts of data and knowledge and are therefore primarily designed for static manufacturing contexts. Hence, to tackle this challenge, state-of-the-art approaches of fault detection are adapted and integrated into the skill-based production. Finally, the elaborated model is prototypically implemented in a realistic demonstration environment to proof the concept.

Since it is difficult to directly model faulty behavior of Cyber-Physical Production Modules (CPPM) in flexible production, their nominal behavior during the execution of atomic skills is modelled and compared with its actual behavior. This procedure additionally brings the benefit of being able to detect a priori unknown faults because no specific fault models are required. The production modules are subdivided to component level to handle behavior complexity and allow different atomic skill combinations. Introducing a component comparing nominal and actual behavior enables the calculation of symptoms based on deviations and defined thresholds. The execution of the developed components is synchronized with skill execution to enable an automated fault detection.

This paper is structured as follows: Sect. 2 shows a current state-of-the-art of skill-based and modular production as well as suitable fault detection approaches. In Sect. 3 the elaborated model components are defined and described. Additionally, their integration into the skill-model is outlined. The prototypical implementation in a real-world demonstrator regarding two scenarios is presented in Sect. 4 whereas finally Sect. 5 gives a conclusion and an outlook for further research.

2 State of the Art

2.1 Modular, Skill-Based Production

Modular production systems are based on the concept of Cyber-Physical Production Systems (CPPS), which consist of several combinable CPPM for realization of a vendor-independent production environment [6]. To build up such CPPS due to a flexible aggregation of lower-level components, like CPPMs, standardized interfaces are necessary to provide the required flexibility to react faster to production changes [7]. CPPMs can provide and encapsulate their manufacturing functions with a common interface by using the skill-based approach, realizing a more flexible combination of CPPMs to CPPS [3]. In [8] an information model for capabilities, skills and services is introduced that defines the corresponding terms as well as describe their relations. Also in [8] a skill is described as an implementation of a function which is deployed on a specific resource, for example in a CPPM. The behavior of a skill follows a standardized state machine and can be accessed by a standardized skill interface [8]. The skills themselves and their underlying functionalities can be modelled following the methodology introduced in [9]. That approach additionally offers an automatic code generation based on the modelled systems and systems group functionalities. In [10] knowledge-based fault diagnosis in skill-based production is introduced covering the following step of the fault detection and diagnosis pipeline after this work.

2.2 Model-Based Fault Detection

Model-Based Approaches (MBA) for fault detection use models to simulate the normal behavior of a plant or product features. The simulated behavior is then compared to observations representing the systems actual behavior to determine symptoms leading to potential faults [11, 12]. In the following an overview about the state of the art is given.

Cazes and Kalech introduce a MBA modelling system components, observations, system description and the components behavior. Additionally, the model is suited for uncertainty over given observations. Consistency-based and abductive models are used to define the scenario and to explain the behavior [13]. In [12] and [14] the nominal dynamic behavior is modelled with state space models with differential equations. The implementation of the observer-pattern uses satisfiability theory modulo linear arithmetic. The observer pattern is used to calculate residuals [12, 14]. Borutzky uses hybrid bond graphs to provide numerical values of unknown parameter degradation functions. Based on these values the remaining useful lifetime is calculated data-based [15]. A structural approach used for detection and isolation of faults in hybrid systems that calculates residuals by defining hybrid minimal structurally overdetermined sets selection problem is shown in [16]. In [17] a MBA is introduced that uses machine learning techniques to learn the nominal behavior of CPPS. The created residuals are fed into a diagnosis model for root cause analysis.

Several MBA are proposed, each focusing on a limited application. In recent research skill-based production is introduced to enable more flexibility. Some papers address

CPPM but there is still no explicit integration into the skill-based production approach. This paper fills the gap of introducing a model-based methodology to model fault detection especially for skill-based production and shows an integration for automated usage.

3 Integration of Fault Detection into Flexible Skill-Based Production Systems

The main task of fault detection is the determination of faults present in a system. The identification and classification of faults is not part of the fault detection in this work. Here, the system is defined as a CPPM or Cyber-Physical-System (CPS) that can execute at least one skill and act in a flexible production environment.

Depending on the chosen approach, symptom generation can be used to classify a system's behavior as anomalous or normal. For the symptom generation the phenomenological and model-based approach can be distinguished. Phenomenological approaches usually use expert system or data-based system to classify system observations directly as anomalous or correct. In expert systems the expert's knowledge is modelled manually to be used for classification decisions whereas data-based approaches (e.g. [18, 19]) extract and model their knowledge automatically from historic data. The higher automation and flexibility make data-based approaches more suitable for ever changing production systems.

On the other hand, model-based approaches can be used for an indirect detection of faulty behavior. In those approaches models are used to simulate the nominal behavior of a production system or normal product features. Nominal models need relevant input to enable the calculation of the features representing nominal behavior. The simulation results are then compared to the observations of the system. If the deviation between the values is significantly a symptom for a potential fault is detected [11].

Since faults are rare compared to normal behavior even in conventional production systems it is even more complicated to get enough data or knowledge for phenomenological approaches in flexible production systems. Therefore, this paper uses model-based approaches to bring fault detection to skill-based production systems, whereas faulty behavior is defined as the deviation of at least one feature from nominal behavior.

3.1 Required Components for Fault Detection in Skill-Based Production

For building a system to automatically generate symptoms and thus detect faults in flexible skill-based production systems we identified three main components: Nominal models, system observation and the deviation-based symptom generation. The following subsections describe those required components and their integration into the skill model.

Nominal Models in Skill-Based Production: The task of a nominal model is to describe and simulate the normal behavior of a defined system. In this work the system taken into consideration is a CPS or CPPM that can provide at least one skill. Nominal models therefor highly depend on relevant input parameters to achieve a precise quality in calculation of the nominal behavior, e.g., product features, status of the CPPM or parameter sets of executed skills.

Depending on the specific skill and the resource on which the skill is to be executed, a minimum set of features is identified that is required to describe the nominal behavior of the resource during the execution of the skill. To get a better and fuller picture of the behavior description the feature set can be extended when necessary.

The identified features need to be described by nominal models. The selection of a suitable model depends on the resource, the skill, the feature itself, the available data quantity and quality and available experts to build the models. Thus, the models can be data-based, statistical models, model-based like observer-based models, dynamic or state space models as well as parity equations, first principle models or defined curves from experience e.g.

System Observation in Skill-Based Production: For the observation of the system a set of defined features during the execution of a skill is monitored, whereas the system and the set of features need to match with the definitions for the nominal models. The selected feature set and their measurements display the actual behavior of the system.

For enabling the monitoring of the features, a set of sensors needs to be installed that can measure the features directly (e.g., temperature) or building the basis for the calculation of the feature (encoder degree to calculate positions) or the simulated or calculated value by the nominal model.

The kind and quality of the sensor as well as the frequency of the sensor measurements need to be specified according to the requirements of the specific feature, the resource and the executed skill depending on the product requirements.

Symptom Generation: To generate symptoms values from the nominal model and corresponding observed values needs to be compared. Based on defined thresholds a deviation can be mapped to a specific symptom. The thresholds can be either set manually by experts or automatically calculated and adapted. The automatically adapted threshold values can for example be learned by machine learning approaches. Depending on different ranges of the thresholds and their direction the symptoms are then labelled as low, very low, high, and very high. The notation is done using plusses and minuses and can be extended depending on the situation.

A detected symptom does not necessarily lead to a fault. If a set of symptoms is identified that is not considered normal behavior but does not lead to a fault a new normal situation is identified that is not covered by the existing nominal model and needs to be added to the nominal model.

Symptoms identified as leading to faults are sent to a fault diagnosis component to be mapped to a fault class.

3.2 Integration of Identified Components in Skill-Model

In this section the integration of the identified components required for a fault detection system, namely the nominal model, the system observation component as well as the symptom generation component, into the skill-model is discussed. The symptom generation component does not require an integration into the skill model and is thus not further elaborated.

The integration involves deploying the nominal model and integrating the monitoring features with the skill-state-machine execution to make them part of the skill model itself. In contrast to that the symptom generation may be an external component that functions outside the scope of the skill-model.

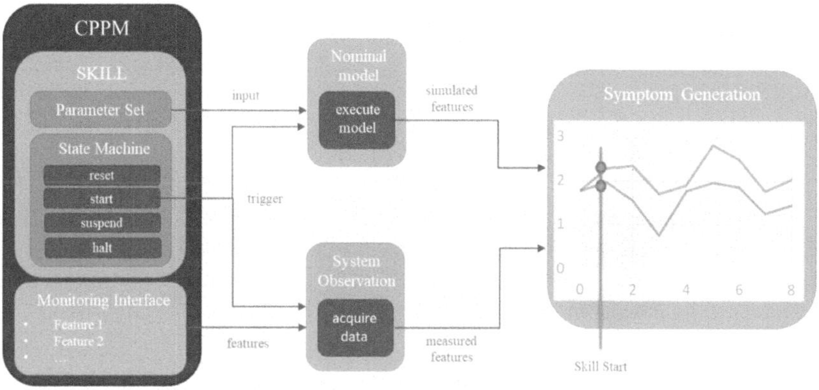

Fig. 2. Overview over required fault detection components

Nominal Model Deployment: The nominal model needs to be integrated into the skill-state-machine. This means it is necessary to couple the transitions defined in the skill model and the nominal model execution. Additionally, the parameter set of a skill defining its input is used as input for the nominal model as well. This ensures that the nominal model is synchronized with the actual skill behavior and accurately represents the expected behavior during skill execution.

Monitoring Component: The monitoring component, as described in Sect. 3.1, observes a defined set of relevant features for every skill and provides the data to the external symptom generation component. To synchronize the actual skill behavior with the acquired data, it is necessary to couple the monitoring component into the skill-state-machine. Additionally, this assures that the data coming from the nominal model as well as the data coming from the monitoring component are aligned, thus allowing the symptom generation component to compare corresponding values. To achieve a sufficient level of detail while simultaneously maintaining a reasonable amount of data the monitoring frequency of the observed parameters should be adjustable according to the requirements of the features.

Once data has been acquired by observing the relevant features of the production system it is stored either locally in a suitable format, e.g., a comma-separated values (CSV) file, or alternatively it is pushed to a database to allow for easy access by multiple users or subsequent processes.

4 Implementation and Results

The proof of concept in this work is implemented on a CPPS of the SmartFactoryKL shared-production ecosystem. The CPPS consists of several CPPM providing different skills to be able to produce product in small batch sizes and is shown in Fig. 3.

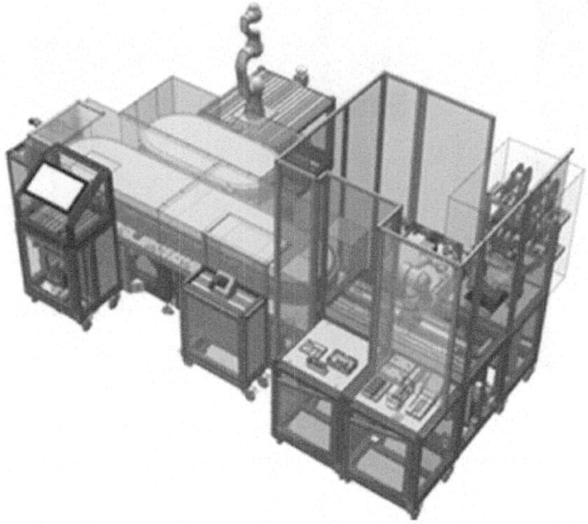

Fig. 3. CPPS of the SmartFactory-KL with modular module CPPM

The considered CPPM in the implementation works as modular module with several submodules. The modular module consists of a 3D-printing station, a central station with robot arm for transporting printed parts to the transportation unit and a manual assembly station. For implementation, the robot arm is considered as Gripper Component (GC), which is provided by a Universal Robots UR5e. All skills provided by the CPPM are controlled and modelled using OPC UA. During the execution of the Pick & Place skill of the GC the energy consumption is considered as an example for monitoring during the skill execution. The fault free condition considers the transport with the load of a 3D printed nubstone truck trailer by the GC. To simulate a faulty behavior due to the selection of a wrong part for the production process, a metal square having almost the dimensions but a much higher weight than the 3D printed product is used.

The nominal model for the fault detection in the use case is based on the energy load profile of the GC, in which the intended load curve during the skill execution for the 3D printed part is described. For the methodical creation of the load profile the load profile clustering from [20] was applied. For the load profile creation of the GC for fault detection, the skill-level is considered. The load profile clustering has been applied and implemented for the GC as shown in [21], to create and monitor energy load profiles during the skill execution. The energy load profile can be used as nominal model and be compared with actual values for a subsequent symptom generation.

Figure 4 demonstrates the comparison of a monitored energy load profile with the nominal model for the expected 3D printed part, which should be transferred to the transport system. The load profile for the expected part is shown as nominal model and compared with the load profile measurements for the monitored GC skill execution. The execution of the Pick and Place-Skill handles several movement sequences which comes with related energy consumption patterns.

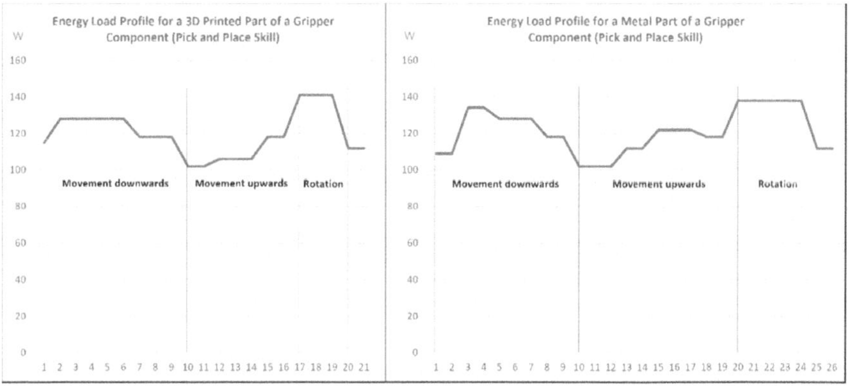

Fig. 4. Monitored energy load profile in comparison with the nominal model of a gripper component

The skill in the implemented and validated example can be differentiated in three parts of movement. At first, the GC starts with a movement downwards to pick the intended part. Second, the gripper picks the part and moves upwards. From this movement sequence on, the duration of the skill execution as well as the energy consumption can vary for the first time due to the part weight. At least, the gripper moves with a rotation towards the transport system, to place the picked part on a workpiece carrier. Figure 4 shows, how these three movement sequences are mapped to patterns of the related energy load profile for usage as nominal model as well as for monitoring purpose and comparison with the nominal model during the skill execution. While the energy consumption of the first movement sequence is nearly identical since the part weight does not matter, a higher energy consumption can be observed in the energy load pattern during the upwards movement and in the rotating movement with the loaded metal part. The metal part with the higher weight also results in a longer execution time of the second and last movement sequence in comparison with the 3D printed part. Due to the differences of the patterns of the energy load profiles during skill execution, a different energy consumption can be observed in total. While the skill execution for the 3D printed part requires 0,667 Wh in total, it is 0,844 Wh for the metal part, which is mainly explained by the longer skill execution time. The values of this energy load profile comparison for both parts are shown in Table 1.

Within the symptom generation component, a deviation for the movement upwards and the rotation of the GC was calculated exceeding the threshold of 0,1 Wh set by process experts. The deviation thus leads to a symptom of the feature energy indicating to a fault picking up a wrong product. To get a classification and analysis of the potential

Table 1. Data of the gripper component for a Pick and Place-Skill with a 3D printed part and a metal part

Movement Sequence	Duration (s)		Energy Consumption (Wh)	
	3D Printed Part	Metal Part	3D Printed Part	Metal Part
Movement downwards	9	9	0,308	0,307
Movement upwards	7	10	0,210	0,314
Rotation	4	6	0,149	0,223
Summarized	20	25	0,667	0,844

fault additional symptoms need to be considered and analyzed by a fault diagnosis component e.g. described in [10] that is out of scope for this work.

5 Discussion and Conclusion

In this paper model-based fault detection is integrated into the skill-based production approach for flexible production environments. After elaborating the principle of MBA and the benefit of detection of a priori unknown fault that are especially interesting in flexible, ever-changing production systems the required components nominal models, monitoring, and symptom generation are outlined in the context of skill-based production environments.

Although existing approaches tackle fault detection for CPPM, no approach explicitly addresses the integration in the skill-based approach. Nominal models define the normal behavior of a resource executing a skill including a specific parameter set. Using the level of skills for the description of the nominal behavior reduces the model complexity due to the division into smaller parts compared to a model describing the whole resource. The features that need to be monitored are identified for a skill specifically. Thus, monitoring only exactly the required features for the execution of a skill reduces unused data of skill-unrelated sensors.

The integration into the skill model in terms of synchronization is shown enabling an automatic execution of the fault detection components. The parameter set describing the skill as well as the monitored features of the monitoring interface (see Fig. 2) are used as input for the nominal model and the system observation respectively. The synchronization between the execution of the skill and the fault detection is implemented using a trigger of the skill state machine.

Finally, a prototypical implementation displaying the principle of the concept is demonstrated at a real-world demonstrator of SmartFactoryKL. Therefore, energy load profiles are recorded during the fault free execution of a pick & place skill used as a nominal model. Real measures of the required energy are compared to the nominal load profile during the transport of a wrong product that is much heavier. Thus, a symptom indicating the pick-up of the wrong product could be generated.

In future research the following additional challenges on this topic can be tackled. Bringing the model complexity to the skill level comes with the challenge of combining the behavior of multiple skills executed simultaneously and thus, the modelling of those combined behaviors. The threshold values for deviations can be dynamically adapted to the context and environmental conditions in the production systems. Furthermore, more sophisticated models can be used and tested in more complex environments. For more autonomy of CPPM an agent-based system can be developed enabling automated communication especially between fault detection and diagnosis components.

References

1. Hermann, J., Rübel, P., Birtel, M., Mohr, F., Wagner, A., Ruskowski, M.: Self-description of cyber-physical production modules for a product-driven manufacturing system. Procedia Manufact. **38**, 291–298 (2019)
2. Kagermann, H., Wahlster, W., Helbig, J.: Recommendations for implementing the strategic initiative INDUSTRIE 4.0. Final report of the Industrie 4.0 Working Group. Forschungsunion, Berlin (2013)
3. Ruskowski, M., et al.: Production Bots für Production Level 4. atp **62**, 62–71 (2020)
4. Wilhelm, Y., Reimann, P., Gauchel, W., Mitschang, B.: Overview on hybrid approaches to fault detection and diagnosis: combining data-driven, physics-based and knowledge-based models. Procedia CIRP **99**, 278–283 (2021)
5. Isermann, R., Ballé, P.: Trends in the application of model-based fault detection and diagnosis of technical processes. Control. Eng. Pract.Pract. **5**, 709–719 (1997)
6. Kolberg, D., Hermann, J., Mohr, F.: SmartFactoryKL Systemarchitektur für Industrie 4.0-Produktionsanlagen. Whitepaper, Kaiserslautern (2018)
7. Birtel, M., Mohr, F., Hermann, J., Bertram, P., Ruskowski, M.: Requirements for a human-centered condition monitoring in modular production environments. IFAC-PapersOnLine **51**, 909–914 (2018)
8. Köcher, A., et al.: A reference model for common understanding of capabilities and skills in manufacturing. at - Automatisierungstechnik **71**, 94–104 (2023)
9. Köcher, A., Hayward, A., Fay, A.: Model-based engineering of CPPS functions and code generation for skills. In: 2022 IEEE 5th International Conference, pp. 1–8 (2022)
10. Rübel, P., Moarefvand, N., Motsch, W., Wagner, A., Ruskowski, M.: Enabling fault diagnosis in skill-based production environments. In: 2023 IEEE 28th International Conference on Emerging Technologies and Factory Automation (ETFA). 2023 IEEE 28th International Conference on Emerging Technologies and Factory Automation (ETFA), Sinaia, Romania, 12.09.2023 - 15.09.2023, pp. 1–8. IEEE (2023)
11. Niggemann, O., Lohweg, V.: On the diagnosis of cyber-physical production systems. In: Proceedings of the AAAI Conference on Artificial Intelligence, pp. 4119–4126 (2015)
12. Diedrich, A., Maier, A., Niggemann, O.: Model-based diagnosis of hybrid systems using satisfiability modulo theory. AAAI **33**, 1452–1459 (2019)
13. Cazes, D., Kalech, M.: Model-based diagnosis with uncertain observations. AAAI **34**, 2766–2773 (2020)
14. Diedrich, A., Niggemann, O.: Diagnosing hybrid cyber-physical systems using state-space models and satisfiability modulo theory (2018)
15. Borutzky, W.: A hybrid bond graph model-based - data driven method for failure prognostic. Procedia Manufact. **42**, 188–196 (2020)
16. Khorasgani, H., Biswas, G.: Structural fault detection and isolation in hybrid systems. IEEE Trans. Automat. Sci. Eng. **15**, 1585–1599 (2018)
17. Bunte, A., Stein, B., Niggemann, O.: Model-based diagnosis for cyber-physical production systems based on machine learning and residual-based diagnosis models. AAAI **33**, 2727–2735 (2019)

18. Bampoula, X., Siaterlis, G., Nikolakis, N., Alexopoulos, K.: A deep learning model for predictive maintenance in cyber-physical production systems using LSTM autoencoders. Sensors **21**, 972 (2021)
19. Stavropoulos, P., Papacharalampopoulos, A., Sampatakakis, K.: A CPS platform oriented for quality assessment in welding. MATEC Web Conf. **318**, 1030 (2020)
20. Teiwes, H., Blume, S., Herrmann, C., Rössinger, M., Thiede, S.: Energy load profile analysis on machine level. Procedia CIRP **69**, 271–276 (2018)
21. Motsch, W., et al.: Energy agents for energy load profiling in modular skill-based production environments. In: Borangiu, T., Trentesaux, D., Leitão, P., Berrah, L., Jimenez, J.F. (eds.) Service-Oriented, Holonic and Multi-Agent Manufacturing Systems for Industry of the Future (SOHOMA2023), pp. 394–408. Springer, Cham (2023). https://doi.org/10.1007/978-3-031-53445-4_33

Towards an Advanced Artificial Intelligence Architecture Through Asset Administration Shell and Industrial Data Spaces

Michel Iñigo[1(✉)], Jon Legaristi[2], Felix Larrinaga[2], Ekhi Zugasti[2], Javier Cuenca[2], Blanca Kremer[3], Daniel Estepa[3], Mikel Ayuso[4], and Elena Montejo[5]

[1] Innovation and Technology Department, MONDRAGON Corporation, Mondragon, Spain
minigo@mondragoncorporation.com
[2] Electronics and Computer Science Department, Mondragon Unibertsitatea, Mondragon, Spain
[3] ICT Department, Ikerlan, Mondragon, Spain
[4] ICT Department, Lortek, Ordizia, Spain
[5] ICT and Automation Department, Ideko, Elgoibar, Spain

Abstract. This article develops an architecture for the implementation of Artificial Intelligence in the manufacturing value chain based on standard technologies and data spaces. The standards considered are IEC 63278 "Asset Administration Shell (AAS) for industrial applications" and DIN SPEC 27070:2020 – "Requirements and reference architecture of a security gateway for the exchange of industry data and services" by IDSA. The architecture provides a data space that allows MONDRAGON industrial cooperatives to use data for the execution of advanced data analytics, Artificial Intelligence (AI) algorithms and interoperability between assets and IoT-platforms. The development of knowledge in this field allows, on the one hand, to optimise the consolidation of data as a strategic factor and, on the other hand, to increase collaboration between manufacturing companies, suppliers and technology providers. The article also explores specific Artificial Intelligence technologies with a wide application in industrial environments. In particular, the study has focused on research into Low/No Code, Explainability (XAI) tools and incremental learning algorithms. The contributions of this paper are summarised in 1) creating an IDS-AAS based architecture and data space that allows the exploitation of AI use cases, either by directly downloading models or by using AI as a service, 2) identifying useful AI tools for industry such as AutoML, No/Low code, XAI or incremental learning, 3) implementing a use case where different AI use alternatives are implemented.

Keywords: AI applications in manufacturing systems · Data spaces · Asset Administration Shell · Digital platforms

1 Introduction

Current industrial ecosystems integrated in value chains must operate effectively in very dynamic environments due to complex and distributed production and increasing incorporation of digital technologies in industrial systems. To achieve it, advance analytics

A. Wagner et al. (Eds.): ESAIM 2023, LNME, pp. 35–42, 2024.
https://doi.org/10.1007/978-3-031-57496-2_4

are necessary. In particular, Artificial Intelligence and the interconnection and interoperability between stakeholders and digital platforms in the value chain is a necessity where suppliers and consumers exchange their assets data. Within Data Space arises as a technology that allows data to be connected securely and efficiently in an associated and distributed production within a value chain [1].

Consequently, high interest and necessity in data generation and management has highlighted a crucial issue about the lack of standardized models, structures and technologies to share these data, ensure interoperability and sovereignty. To carry out these concepts Asset Administration Shell (IEC-63278), GAIA-X and IDS (DIN SPEC 27070:2020) standards have emerged as part of Manufacturing Data Spaces essential to build interoperable, autonomous data ecosystems quickly and efficiently throughout all European Industry [2], meeting the requirements of industrial products, services and processes complying with reference models and the I4.0 language. However, there is a lack of jointly implementation in a dynamic ecosystem dominated by small and medium-sized enterprises, where a risk of becoming dependent on monopoly platform operators is a reality.

In this I4.0 context, MONDRAGON Corporation, which manufactures different products in the industrial value chain, such as aesthetic, interior and bodywork automotive components, CNC or machine tools, to name but a few, has participated in previous EU initiatives and projects such as QU4LITY, Arrowhead or more recently Flex4Res. All to boost digital transformation and to prepare their industrial and technology providers to Manufacturing Data Spaces challenges involving specifically AAS and IDS as can be seen in [3] and [4]. In addition, Artificial Intelligence services, specifically XAI, No/Low-Code should be provided between technology providers and industrial consumers in a simplified, fast and trusted manner. The aim of this paper is to set up an architecture to respond to the challenges of implementing a use case for AI services in a collaborative industrial environment with a Industrial Data Space. The use of AAS enables a manufacturing namespace and IDS allows to interchange data and standardized AI models for AI services.

2 Background

2.1 Artificial Intelligence

Artificial Intelligence is an increasingly essential part of any industry and sector. The use of AI models has gone from being an almost exclusive asset of large technology companies to being present in almost any activity we do in our day-to-day lives. Industry is no exception to this. Over the next few years, those companies that manage to incorporate the advantages of artificial intelligence more efficiently into their production chain will have a strategic advantage over their competitors. But when we talk about AI we are actually talking about a very diverse set of technologies with clearly differentiated objectives, some of them being very relevant for industrial use cases and others not. For this reason, we consider that some fields of application are of particular relevance for the industrial field, highlighting the following:

- **Low-Code/No-code technologies**: Also known as AutoML. They offer automatic ways to create ML models, with littlehuman interaction, either using very simple code (Low-Code) or no code at all (No-Code). Some interesting tools referring to this field are Tpot, autoViML, pycaret, dataiku or orange [5].
- **Model Explainability (XAI):** A set within AI that has as its basis the explainability of AI models. XAI solutions for anomaly detection models in the digital twin use case is one of the most interesting field of application of this technology. Intrinsic explanations [6], Attribution based methods, Counterfactual explanatios [7], Gradient based methods for time series are some of the most relevant techniques.
- **Incremental learning**: incremental learning techniques allow the creation of AI models that evolve over time. In this way, instead of needing to have all the data available at once to train the model, it can improve its performance with the arrival of new information, which is very suitable for industrial environments [8].

2.2 International Data Spaces

In recent years the main players in the European industry have embraced the IDS standard, created by the Industrial Data Space Association (IDSA), to design an environment that allows different organizations to create shared data spaces which enable decentralized data sharing. Each organization can use third party data and knowledge to improve their own productive and operative processes, as well as share and monetize their own data.

The IDS is a virtual data space that uses existing technologies and standards, as well as extensively used governance models, to enable secure and standardized data exchange in a trusted ecosystem. It provides the core elements necessary to create smart-service industrial scenarios while at the same time guaranteeing data sovereignty for data owners [9]. The main goal of the IDSA is ensure data sovereignty, which can be defined as the capability of being entirely self-determined regarding data [9, 10]. With this main objective in mind, the IDSA proposes the IDS Reference Architecture Model (RAM), the main requirements that the model includes to ensure secure and trusted data exchange are the following: Data sovereignty and security, trust, ecosystem of data, standardized interoperability, value adding apps and data markets.

The amount of data shared in these collaborative virtual spaces is too large to be handled by human negotiations, it must be negotiated, executed and monitored semi automatically for contractual and legal compliance [4]. Most of these decisions are not trivial and many times they require some kind of interaction between the system and the user for policy compliance. The system should allow the execution of all necessary actions, some of which are: Capture, store, enrich and distribute data, determine and enforce data usage conditions, ensure data quality, monitor and log data transactions [11].

2.3 Asset Administration Shell

The Asset Administration Shell (AAS) is a key component of the Industrie 4.0 vision. The AAS, based on IEC 62832 - Digital Factory Framework and defined in IEC 63278, is the implementation of the digital twin (DT) of an asset. It consists of a number of submodels that describe all the information and functionalities in a uniform and standard data language.

The 2030 Vision for Industrie 4.0 [12] defines three strategic fields of action that are crucial for a successful implementation of Industrie 4.0: autonomy, interoperability and sustainability. The interoperability of all stakeholders in complex and decentralized systems is a key strategic element. The AAS fulfills the needed requirements [10] to achieve the necessary interoperability through its ability to provide a standardized and uniform way to represent and exchange information about assets and I4.0 components between partners across companies in the value chain.

Manufacturing companies are increasingly adopting AI in their facilities, empowered by digital transformation. The interconnection of various types of machine and user data in a well-integrated ecosystem based on a standardised architecture with open standards is a real need [13]. AAS can be used to improve asset interoperability for AI assets and thereby ensure sustainable reliability. The AI assets can be seamlessly integrated into existing systems and processes, reducing the risk of incompatibilities or other issues that could impact the reliability of the system as a whole, thus enabling the development of new solutions and business models.

3 Advanced AI Architecture

The AI platform implemented is based on the open-source components [14] provided by IDSA (International Data Space Alliance). Figure 1 presents the infrastructure of the instantiated dataspace. The dataspace follows the FAIR (Findable Accessible Interoperable and Reusable) principles for good data governance. The enabling components of the shared data space ecosystem are presented next.

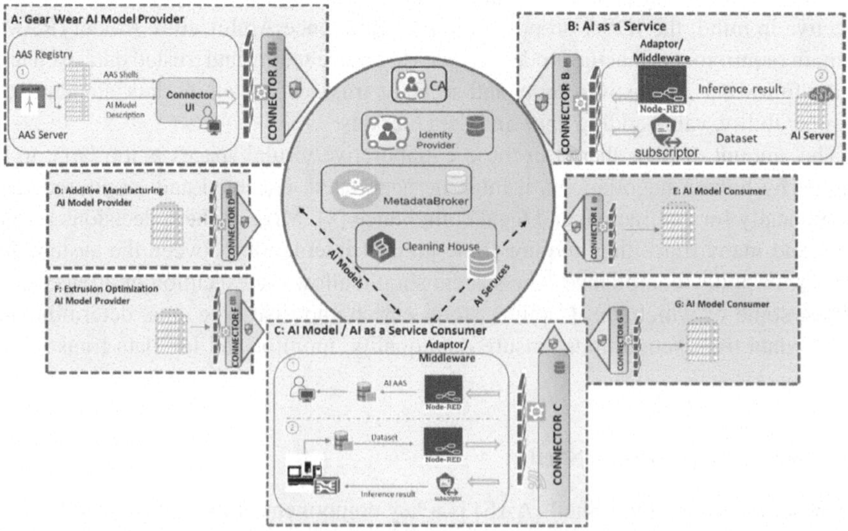

Fig. 1. Advance AI architecture with AAS-IDS

- **CA**: The Certification Authority is in charge of generating the credentials for the participants in the data space.
- **IdentityProvider**: This component is in charge of verifying the identity of the participants and generating temporary tokens that allow secure communication between participants. The component is responsible for the authentication of participants in any data exchange within the data space.
- **MetadataBroker**: This component enables the discovery of catalogues and resources offered by the participants. Each participant will register in this component its connector or connectors and the resources offered by these components with metadata that will make it easier for another participant to find those resources and the connector that provides them.
- **ClearingHouse**: This component is in charge of recording all transactions between providers and consumers, as well as the contracts and rules on which these transactions are based.
- **Connector**: This component acts as an interface between the data to be shared and the data space. It encapsulates all communications between participants in a secure and interoperable way. It does this by providing a REST API that enables other participant-specific software components to interact with the data space. Through this REST API, a data provider participant can register the connector in the Metadatabroker and expose internal data of interest as connector resources and register them in the Metadatabroker. Similarly, a consuming participant can register a connector in the ecosystem, select a resource of interest by discovering it in the Metadatabroker and consume it directly from the provider offering it. The connector also implements subscription mechanisms to resources that allow a participant to receive new data if the provider updates the resource. A connector can be both a provider and a consumer of data. The specification of a common REST API for all connectors within the data space addresses the interoperability aspects of FAIR data. This component is key to interoperability because it allows data to be offered and accessed in a standardized way, without worrying about the technologies that each participant may use to manage and store its data internally.
- **Connector Adapter**: This is a middleware required to manage resource management of the connectors and update content through the IDS connector subscription system. For the AI as a Service model, the middleware is responsible for updating the connector resources necessary to be able to perform the inference and transfer of data. In addition, this middleware plays the role of IoT Gateway to process and store a large amount of data and to inference knowledge that can be generated by different assets in an industrial plant. The middleware is also able to provide support for most connectivity protocols (CoAP, MQTT, REST, OPC UA). Through the IDS connector subscription system, this middleware offers a layer of REST services to manage notifications and to be able to carry out the entire flow from the moment a dataset is sent to infer a result, until the results of the AI model prediction are returned [4].
- **AI Server**: This is a server used to deploy the model and enable inference through a REST API.

- **Asset Administration Shell repository**: This component is necessary to register the AI submodels defined in the different use cases. The Eclipse BaSyx AAS server has been used, which allows, on the one hand, to upload both AAS instances and submodels to the repository and, on the other hand, to access through a REST layer the AAS resources stored in the server.

Based on this architecture two use cases have been envisioned. On one side the sharing of AI models and on the other the execution of inference models as a service.

3.1 Use Case 1: Sharing Machine Learning Models in a Data Space

The idea behind this manufacturing scenario is to be able to create value chains between suppliers of machine learning models and potential consumers of these models. Once an AI model has been trained to solve a certain type of problem, from optimizing or detecting faults in a production process to predict maintenance needs, this model could potentially be used by different companies that have similar needs but not the capacity to train their own models.

An important challenge to achieve this scenario is to be able to describe the characteristics of any machine learning model in the most standard way possible. In our case, we have worked on categorizing the aspects that define an artificial intelligence model and we have represented this categorization as a submodel of AAS. This enables consumers to access this resource in a standardized way and to analyze whether the model registered in the data space is adequate or not to solve their problems.

In our proof of concept, different inference model providers are registered as participants in the Mondragon Data Space (green boxes in Fig. 1) and each provider offers its trained model. For each model, two resources are registered in the data space, one of the resources is the AAS submodel, uploaded to the AAS repository, of the model with all the features of how it has been trained and tested together with its validation metrics. The other resource is a compressed folder with everything needed to run the inference engine.

Each provider deploys an IDS connector in its digital infrastructure and registers itself as participant in the Mondragon Data Space. Once registered, using the middleware it registers the two resources associated with the model together with the usage and payment policies for each of them in the Metadatabroker. Certain metadata are added to allow the discovery of the resources and to facilitate their selection by potential consumers.

Each Consumer (grey boxes in Fig. 1) deploys another IDS connector in its digital infrastructure and registers itself in the same data space. Consumer, using the middleware, queries the Metadatabroker to search for providers that can offer inference models of interest. The consumer, making use of the information provided by the Metadatabroker, selects the model of interest and obtains the url to access the resource containing the features of the model. From now on, consumer, once the contract agreement has been signed, is able to download the AAS submodel,, analyses the details (type of algorithm, libraries used, necessary preprocessing, input data schema, etc.,) and decides whether it is suitable for its purpose. If so, consumer will download the inference engine and run it on its premises using its own data.

3.2 Use Case 2: Model Inference Execution Result as a Resource in a Data Space

In this use case we consider a scenario where value chains are created between on the one hand participants that provide the results of the execution of AI models that require high computational capabilities and on the other hand participants that provide the input data and benefit from the results of the execution.

In this proof of concept, we have two participants registered in the data space. Participant B (orange box in Fig. 1) is a provider capable of running AI models on its central server to detect maintenance needs. Participant B offers the execution result as a resource through its connector. Participant C (light grey box in Fig. 1) is a production company that is interested in a service that based on the daily data provided by its machines gets a report on maintenance actions to be taken into account the next day. Participant C offers this data as a resource through its connector.

The role of each participant is both provider and consumer. On the one hand, participant B subscribes to the machine data resource provided by participant C and on the other hand participant C subscribes to the resource that participant B provides with the result of the inference execution. Daily the production company collects data from its machines and updates the resource in its connector, the AI service provider receives the data modification event and runs the inference engine with the new data. Once the execution is finished, the resource with the result is updated and an event is sent to the company with a report about.maintenance actions.

4 Conclusions and Future Steps

Combining the current state of the AAS and International Data Spaces technologies can be used to deploy an Industrial Collaborative Data Spaces which AI services are carried out interchanging AI models or algorithms as a Service. The architecture defined enables on the one hand common building blocks such as CA, Identity Broker, Metadata Broker and Clearing House and on the other hand providing connectivity and interoperability between providers and consumers. Additive Manufacturing, Extrusion Process Optimization and Gear AI models have been shared through MONDRAGON data space as providers. Registering AI model as a resource within a data space improves the competitive capacity of suppliers by being able to reach a larger number of customers in a secure and sovereign way. On the other hand, it allows companies to access a trusted ecosystem in an interoperable way. Although, the possibilities of the AAS and IDS towards interoperability and standardized in a collaborative data space environment are guaranteed, automatic operations should be carried out to agile Edge-Cloud information flows in a whole manufacturing ecosystem together with technology stakeholders. Our next step, will be to expose more advanced collaborative scenarios as well as explain in a deeper way AI models described in AAS. Regarding architecture, MONDRAGON data space modules should be updated or extended ensuring data spaces compliance promoted by IDSA and Asset Administration Shell standard. Besides, it should be extended with modules that allow retraining AI models considering stakeholders data. Finally, data sovereignty and transaction control must continue to be key.

References

1. Quadrini, W., Cimino, C., Abdel-Aty, T.A., Fumagalli, L., Rovere, D.: Asset administration Shell as an interoperable enabler of Industry 4.0 software architectures: a case study. Procedia Comput. Sci. **217**, 1794–1802 (2023)
2. Otto, B., ten Hompel, M., Wrobel, S.: Designing Data Spaces: The Ecosystem Approach to Competitive Advantage, p. 580. Springer, Cham (2022). https://doi.org/10.1007/978-3-030-93975-5
3. Inigo, M.A., Porto, A., Kremer, B., Perez, A., Larrinaga, F., Cuenca, J.: Towards an asset administration shell scenario: a use case for interoperability and standardization in industry 4.0. In: NOMS 2020–2020 IEEE/IFIP Network Operations and Management Symposium, pp. 1–6. IEEE, April 2020
4. Larrinaga, F.: Data sovereignty for AI pipelines: lessons learned from an industrial project at mondragon corporation (2022)
5. Ogundare, O., Araya, G.Q., Qamsane, Y.: No code AI: automatic generation of function block diagrams from documentation and associated heuristic for context-aware ML algorithm training. In: 2022 7th International Conference on Mechanical Engineering and Robotics Research (ICMERR), pp. 191–195. IEEE, December 2022
6. Labaien, J., Idé, T., Chen, P.Y., Zugasti, E., De Carlos, X.: Diagnostic spatio-temporal transformer with faithful encoding. Knowl.-Based Syst. **274**, 110639 (2023). https://doi.org/10.1016/j.knosys.2023.110639
7. Labaien Soto, J., Zugasti Uriguen, E., De Carlos Garcia, X.: Real-time, model-agnostic and user-driven counterfactual explanations using autoencoders. Appl. Sci. **13**, 2912 (2023). https://doi.org/10.3390/app13052912
8. Ade, R.R., Deshmukh, P.R.: Methods for incremental learning: a survey. Int. J. Data Min. Knowl. Manage. Process **3**(4), 119 (2013)
9. Otto, B., et al.: IDS RAM 4, IDS-RAM 4 (2023). https://docs.internationaldataspaces.org/ids-ram-4/. Accessed 18 June 2023
10. Jarke, M., Otto, B., Ram, S.: Data sovereignty and data space ecosystems. Bus. Inf. Syst. Eng. **61**, 549–550 (2019). https://doi.org/10.1007/s12599-019-00614-2
11. Otto, B., Jarke, M.: Designing a multi-sided data platform: findings from the international data spaces case. Electron Markets **29**, 561–580 (2019). https://doi.org/10.1007/s12525-019-00362-x
12. Plattform Industrie 4.0. The 2030 vision for industrie 4.0: how we can shape the digital ecosystem of tomorrow globally (2021). https://www.plattform-i40.de/IP/Redaktion/EN/Downloads/Publikation/Vision-2030-for-Industrie-4.0.pdf
13. Plattform Industrie 4.0. Interoperability in Industrie 4.0: position paper on interoperability in industrie 4.0 (2019). https://www.plattform-i40.de/IP/Redaktion/EN/Downloads/Publikation/position-paper-interoperability.pdf?__blob=publicationFile&v=1
14. https://github.com/International-Data-Spaces-Association

A Deep Reinforcement Learning Approach for Production Scheduling with the Use of Dispatch Rules

Panagiotis Mavrothalassitis, Emmanouil Bakopoulos, Vasilis Siatras, Nikolaos Nikolakis, and Kosmas Alexopoulos(✉)

Laboratory for Manufacturing Systems & Automation (LMS), Department of Mechanical Engineering & Aeronautics, University of Patras, Rio-Patras, Greece
alexokos@lms.mech.upatras.gr

Abstract. This work proposes a framework of a Deep Reinforcement Learning (DRL) scheduling agent that solves the production scheduling problem by deciding the dispatch rules to be used for tasks allocation to resources. For training, Deep Q-Learning is selected as the learning method and a Deep Q Network (DQN) is used for the core of the agent. A Discrete Event Simulation (DES) model represents the production system and the agent proposes actions in the form of dispatch rules. The agent was trained and tested in a production system of a bicycle industry upon various production scenarios. The DRL scheduling agent proposes production schedules with a 7% less makespan compared to traditional methods, such as shortest processing time (SPT) and longest processing time (LPT) dispatch rules.

Keyword: Artificial Intelligence · Deep Reinforcement Learning · Production Scheduling · Discrete Event Simulation

1 Introduction

Scheduling is an enabler for production efficiency, short time-to-market and costs reduction. This topic has been studied for decades aiming to develop new algorithms for fast and accurate scheduling [1, 2]. Scheduling problems usually encounter an enormous solution space, feasibility constraints, and non-linear behaviour, which makes it extremely challenging to find the optimal solution in a reasonable time [3]. Motivated by these type of problems, Artificial Intelligence (AI) proposes possible solutions by avoiding to exhaustively search all possible alternatives [4].

Dynamic scheduling tends to be more challenging than static scheduling. Data driven approaches, such as Deep Learning (DL) and Deep Reinforcement Learning (DRL), are able to interpret past knowledge, and adapt their decision-making process in order to make accurate predictions for the best possible action in different production states.

This work presents a dynamic scheduling agent, which uses DRL in order to take task allocation decisions. Agent's policy-making behaviour is able to produce accurate scheduling decisions on-the-run by selecting the best possible dispatch rule for each

A. Wagner et al. (Eds.): ESAIM 2023, LNME, pp. 43–50, 2024.
https://doi.org/10.1007/978-3-031-57496-2_5

of the available production resources. A Deep Neural Networks (DNN) is utilized for the optimal dispatch rule selection and a Discrete Event Simulation (DES) model of a production system is utilized for training and testing. Three different DRL models are compared in a real-life industrial case of a bicycle industry.

2 Literature Review

In manufacturing, the term scheduling is referring to the production tasks allocation to production resources [1]. The method of dispatch rules is one of the approaches used for scheduling, as they are simple to implement, and they can deliver good solutions quickly [5]. There are more sophisticated scheduling methods, for example based on smart search [6] or genetic algorithms [7] that can deliver better solutions compared to dispatch rules that however require more time to be computed. The dynamic scheduling problem is subject to various uncertainties including machine failures, arrival of new urgent jobs, the modification of job due times [8]. Much research has been carried out around the general concept of dynamic scheduling [9, 10]. It needs to be noted that the above methods mostly rely on exploring the solution space and still require some time for producing the output. To avoid this computational delay, DL approaches have gained a great attention [11]. To this end, reinforcement learning (RL) methods such as Q-learning are proposed for the selection of the proper dispatching rule for a resource [12]. The following approaches are some key research results in the field: self-organizing and self-learning RL-based system for multi-objective dynamic scheduling [13], continuous state features as input of the Deep Q-Network for the approximation of a state-action value function [14], closed-loop rescheduling strategy for handling unforeseen events [15], Graph Neural Network to solve the job-shop scheduling as the agent can learn dispatch rules from scratch with elementary raw features and demonstrates strong performance against the best existing dispatch rules [16]. In [17] authors proposed a methodology for minimizing reconfiguring actions in a production system. In their work, DRL is implemented to make autonomous decision with a built-in DES model. As a result of self-learning, the agent outperforms conventional First-In-First-Out dispatching rule by completing assigned order lists while minimizing reconfigurations.

To find the best dispatch rule for each resource between the Shortest Process Time (SPT) and Longest Process Time (LPT) at each scheduling decision point, this work uses DRL. The action space is defined by the various combinations of these rules and the available production resources.

3 DRL Scheduling Agent Framework

The DRL agent has been developed to solve the Dynamic Scheduling Problem (DSP). DRL agent is also combined with a Deep Q-Network to approximate a state-action value function. The DES model has been developed with the use of Lanner Witness Horizon [18]. The DRL and DES connection is realized with the use of a custom API and files exchange (Fig. 1). The main components of the architecture are: i) DRL agent utilizing a DQN, ii) Execution controller, supporting the whole workflow operation, iii) Scheduling input, consisting of workload related parameters and iv) Environment, utilizing a DES model and representing the state space.

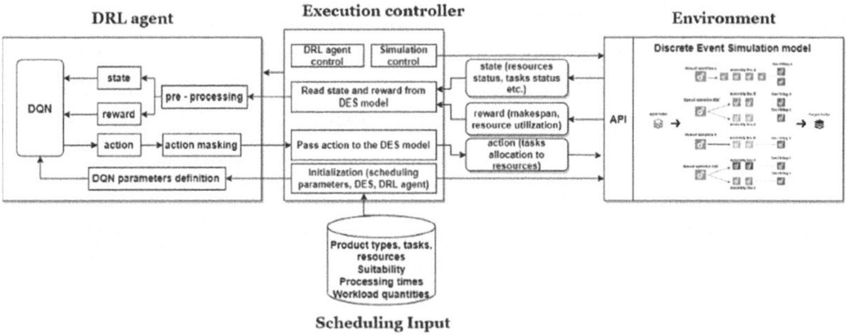

Fig. 1. DRL scheduling framework

3.1 Scheduling Input

The inputs for training and testing the DRL agent are:

- Set of production type IDs is denoted as $Pr = \{Pr_1, Pr_2, \ldots, Pr_{P_{max}}\}$, where P_{max} is the maximum number of production types of a scheduling instance
- Set of suitability $B(i,j)$ where i refers to a task and $0 \leq i \leq N_{max}$ while j refers to a resource and $0 \leq j \leq M_{max}$. M_{max} is the maximum number of resources and N_{max} is the maximum number of tasks of a scheduling problem instance. If a task i can be executed by a resource j then the value is 1, otherwise it is 0
- A set of pre-conditions $C(i, g)$, where i and g refers to tasks for $0 \leq i \leq N_{max}$ and $0 \leq g \leq N_{max}$. If a task g is the precondition of task i then the value is 1 otherwise it is 0
- Each product type has a sequence of tasks that define its process plan. Let us denote this set of tasks as $H(Pr_k)$, where $0 < k \leq P_{max}$
- A set with the quantities that should be scheduled for each product type $Qs = \{Q_1, Q_2, \ldots, Q_{P_{max}}\}$, where $N = \sum_{k=0}^{P_{max}} Q(k) * |H(Pr_k)|$ are the number of tasks and $0 < N \leq N_{max}$
- A list of resource IDs $M = \{M_1, M_2, \ldots, M_{M_{max}}\}$ that exist on the production floor
- The processing times for each task to each resource $A(i,j)$. If task i can be executed by resource j then processing time is greater than 0, else it is 0.

3.2 DRL Agent

The Q-learning is an off-policy temporal difference algorithm and is based on the idea of Q-function [19]. In the following equation the $Q^{\pi}(s_t, a_t)$ is the expected return of the discounted sum of rewards at state s_t by taking action a_t.

$$Q^{\pi}(s_t, a_t) = \max_{\pi} \mathbb{E}\Big[r_{t+1} + \gamma r_{t+2} + \gamma^2 r_{t+3} + \ldots |s_t = s, a_t = a, \pi\Big] \quad (1)$$

The agent in a decision point t in a state $s_t \in S$ selects an action $a_t \in A$ according to a policy π. Taking the action a_t the agent gets to a state s_{t+1} with transition probability $p(s_{t+1}|s_t, a_t) \in P(S \times A \rightarrow S)$ and reward $r_t \in R$. Additionally, γ is a discount factor at each time step t. Also, a is a learning rate, where $0 < a \leq 1$. The objective for the agent is to find the optimal policy π^* which maximizes the expected sum of rewards. The Q-leaning has some limitations when the state space comprises hundreds of variables. For that reason, the DQN concept was used. Coupled RL with DL techniques Q-tables can be replaced with Q-function approximator with weights [20]. In order to solve the DSP problem, due to the fact that the solution space consists of many features, DRL DQN concept was used. Let us denote as $Q(s, a; \theta_i)$ the approximate value using Deep Convolutional Neural Network. Additionally, the θ_i are the weights at iteration i of the Q-Network. The experiences are denoted as $e_t = (s_t, a_t, r_t, s_{t+1})$ with each one stored to a dataset $D_t = \{e_1, \ldots, e_t\}$. Chosen uniformly at random an instance from the pool of stored instances, a Q-learning update is applied of each experience $(s, a, r, s') \sim U(D)$.

$$L_i(\theta_i) = \mathbb{E}_{(s,a,r,s') \sim U(D)}[\left(r + \gamma \max_{a'} Q\left(s', a'; \theta_i^-\right) - Q(s, a; \theta_i)\right)^2] \quad (2)$$

θ_i are the weights of the Q-Network at the iteration i and θ_i^- are the network weights used to compute the target in iteration i. Target network parameters (θ_i^-) are updated with the Q-network parameters every c step, where c is a constant number.

The state is a vector of features that characterizes a given input. In this work the state is the status of the resources (down, busy, available), the status of the tasks (waiting, pending, on-going and finished), and finally the production orders. Moreover, an action describes the dispatch rule that is selected by the DRL agent to propose the task allocation over resources. Lastly, the reward function implemented uses the makespan or the average resource utilization, depending on the DRL agent instance.

4 Case Study

In order to test and evaluate the proposed method an industrial pilot case from the wheels assembly department of a bicycles' production industry has been studied. The agent proposes the best dispatch rules for different production workload and resource availability scenarios. For confidentiality reasons the data presented have been altered, however they are representative of the actual system when seen from the scheduling point of view and where used in order to develop a representative DES model of the real production system. The department consists of four parallel assembly lines. Each line consists of a manual operation resource, two to four assembly resources and one to two tire fitting resources. The department produces 12 types of wheels, with each one of them with a specific suitability on the assembly lines and a processing time. The processing time for a wheel assembly operation varies between 2.2 and 12.8 min. The DES model of the assembly line (Fig. 2) has been utilized for training and testing of the DRL agent.

The scheduling framework presented in Sect. 3 has been utilized for implementing three instances of the DRL agent:

- DRL1: the action space is defined by two options (SPT or LPT) for all assembly lines. The reward is the reverse makespan and it was trained for 2000 episodes.
- DRL2: Same as DRL1 apart from the reward which for DRL2 is the resource utilization.

- DRL3: the action space is defined by sixteen options, as it selects one dispatch rule (SPT or LPT) per assembly line. The reward function is the resource utilization, and it was trained for 8000 episodes.

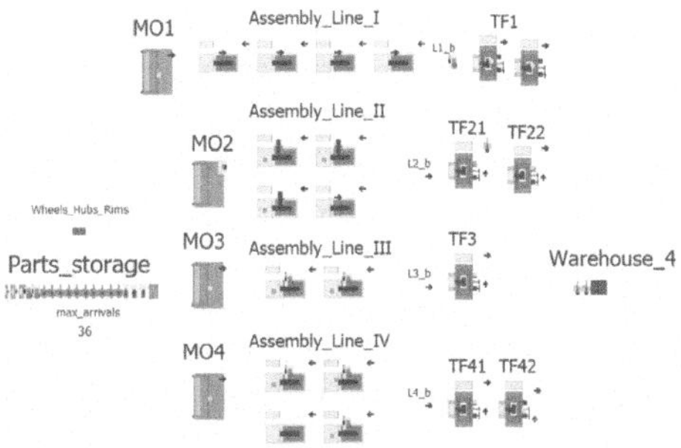

Fig. 2. Wheels assembly department simulation model utilized for training and testing of the DRL agent

For training the DRL agents, various workloads and resources' availability scenarios were generated. A scenario consists of the following two parameters:

- *Workload*: the workload is the quantities for the different wheel types. There are 12 wheel types with a maximum quantity for each type of 12 parts. This means that the maximum workload defines 144 wheels to be produced.
- *Resource availability*: for each scenario, maximum one assembly line is assumed as unavailable (e.g., due to breakdowns).

This variability in the definition of the scenarios, allows the DRL agent to explore several possible instances of the environment of a production system, with various workload combinations and availabilities. For testing and comparing the DRL agents, 100 new scenarios were generated that were not used during the training phase.

5 Results and Comparative Analysis

Comparing the DRL agents' performance with the common dispatch rules (SPT and LPT) the results show that all DRL agent instances outperform the traditional methods, like SPT and LPT. DRL1 achieves in 85% of the testing scenarios a makespan which is equal to the minimum makespan achieved either by SPT or LPT and 15% in which it is equal to the maximum makespan achieved either by SPT or LPT. Similarly, for DRL2 the values are 87% and 13% respectively. DRL1 and DRL2 are limited to achieving equal makespans such the SPT and LPT methods do, since all assembly lines use the same dispatch rule in these instances, but the results show that these agents can select the best rule of the two options in 85% and 87% of the times. However, for DRL3 the dispatch rules are in assembly lines level and are not the same for the whole department. This means that there can be solutions proposed that none of the two dispatch rules could have resulted to. More specifically, DRL3 results in shorter makespan than the minimum makespan achieved either by SPT or LPT for 16% of the testing scenarios. In 66% of the scenarios DRL3 achieves a makespan which is equal to the minimum makespan achieved either by SPT or LPT. In 14% of the scenarios it leads to shorter makespan than the maximum makespan achieved either by SPT or LPT, but greater than the minimum achieved either by SPT or LPT. Lastly, by 4% achieves a makespan equal to the maximum makespan achieved either by SPT or LPT, while there are no scenarios that the agent performs worse than the maximum achieved either by SPT or LPT.

Concluding, it is derived that in average DRL3 performs better than DRL1 and DRL2 (Fig. 3) and results approximately to a 7% shortest makespan than the traditional methods. This is because of their configuration, agents DRL1 and DRL2 propose one dispatch rule for all production lines. Thus, there is a limitation in optimizing as they cannot fine tune their solution by suggesting different rule per line. Moreover, DRL2 is marginally better than DRL1 and DRL3 better than both. The only difference between DRL1 vs. DRL2 is the reward function and this suggests a slightly better performance of the resource utilization reward. However, as the difference between the results of the agents with different reward functions are not significant, no robust conclusion can be drawn on which reward function is better.

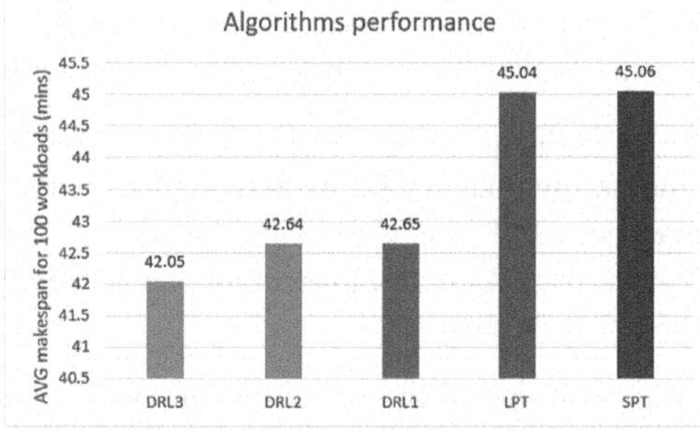

Fig. 3. Scheduling algorithms' performance (average makespan for 100 testing scenarios)

6 Conclusion

In this work, a DRL based scheduling framework is presented in which the DRL scheduling agent uses Deep Q-Network (DQN) for the selection of the best dispatch rules, among SPT and LPT, for a scheduling problem instance. A comparison among three DRL agents' versions and SPT, LPT dispatch rules is reported using data from a bicycle industry. All three DRL agents outperform SPT and LPT in the tests. Moreover, the validation work has shown that the DRL agent can deliver better results when the action space is increased. However, this leads to larger training time, since the capabilities for an action in a certain state increase. Thus, the exploration of the state space lasts longer.

In future work more dispatch rules, such as Most Tasks Remaining per Job (MTRJ) and Earliest Due Date (EDD) will be evaluated. Moreover, investigation for improved reward functions, variations of state or action spaces and potential state of the art implementation capabilities should also be addressed.

Acknowledgments. This work is under the framework of EU Project MAS4AI. This project has received funding from the European Union's Horizon 2020 research and innovation program under grant agreement No 957204. The dissemination of results herein reflects only the authors' view, and the Commission is not responsible for any use that may be made of the information it contains.

References

1. Chryssolouris, G., Subramaniam, V.: Dynamic scheduling of manufacturing job shops using genetic algorithms. J. Intell. Manuf. **12**(3), 281–293 (2001). https://doi.org/10.1023/A:1011253011638/METRICS
2. Ghaleb, M., Zolfagharinia, H., Taghipour, S.: Real-time production scheduling in the Industry-4.0 context: addressing uncertainties in job arrivals and machine breakdowns. Comput. Oper. Res. **123**, 105031 (2020). https://doi.org/10.1016/J.COR.2020.105031
3. Lawler, E.L., Lenstra, J.K., Rinnooy Kan, A.H.G., Shmoys, D.B.: Chapter 9 Sequencing and scheduling: algorithms and complexity. Handbooks Oper. Res. Manage. Sci. **4**(C), 445–522 (1993). https://doi.org/10.1016/S0927-0507(05)80189-6
4. Alexopoulos, K., Sipsas, K., Xanthakis, E., Makris, S., Mourtzis, D.: An industrial Internet of things based platform for context-aware information services in manufacturing. **31**(11), 1111–1123 (2018). https://doi.org/10.1080/0951192X.2018.1500716
5. Papakostas, N.: A scheduling policy for improving tardiness performance. AIJSTPME **2**(3), 79 (2009)
6. Alexopoulos, K., Koukas, S., Boli, N., Mourtzis, D.: Resource planning for the installation of industrial product service systems. IFIP Adv. Inf. Commun. Technol. **514**, 205–213 (2017). https://doi.org/10.1007/978-3-319-66926-7_24/FIGURES/5
7. Wei, H., et al.: Unified multi-objective genetic algorithm for energy efficient job shop scheduling. IEEE Access **9**, 54542–54557 (2021). https://doi.org/10.1109/ACCESS.2021.3070981
8. Chien, C.F., Bin Lan, Y.: Agent-based approach integrating deep reinforcement learning and hybrid genetic algorithm for dynamic scheduling for industry 3.5 smart production. Comput. Ind. Eng. **162**, 107782 (2021). https://doi.org/10.1016/J.CIE.2021.107782
9. Muhamadin, K., Bukkur, M.A., Shukri, M.I., Osama, Elmardi, M.: A review for dynamic scheduling in manufacturing. Type: Double Blind Peer Rev. Int. Res. J. **18**, 25–37 (2018)
10. Wen, X., Lian, X., Qian, Y., Zhang, Y., Wang, H., Li, H.: Dynamic scheduling method for integrated process planning and scheduling problem with machine fault. Robot Comput. Integr. Manuf. **77**, 102334 (2022). https://doi.org/10.1016/J.RCIM.2022.102334
11. Vasilis, S., Nikos, N., Kosmas, A., Dimitris, M.: A toolbox of agents for scheduling the paint shop in bicycle industry. Procedia CIRP **107**, 1156–1161 (2022). https://doi.org/10.1016/J.PROCIR.2022.05.124
12. Wang, Y.C., Usher, J.M.: Learning policies for single machine job dispatching. Robot Comput. Integr. Manuf. **20**(6), 553–562 (2004). https://doi.org/10.1016/J.RCIM.2004.07.003
13. Zhou, T., Tang, D., Zhu, H., Wang, L.: Reinforcement learning with composite rewards for production scheduling in a smart factory. IEEE Access **9**, 752–766 (2021). https://doi.org/10.1109/ACCESS.2020.3046784
14. Hu, L., Liu, Z., Hu, W., Wang, Y., Tan, J., Wu, F.: Petri-net-based dynamic scheduling of flexible manufacturing system via deep reinforcement learning with graph convolutional network. J. Manuf. Syst. **55**, 1–14 (2020). https://doi.org/10.1016/J.JMSY.2020.02.004

15. Palombarini, J.A., Martinez, E.C.: Closed-loop rescheduling using deep reinforcement learning. IFAC-PapersOnLine **52**(1), 231–236 (2019). https://doi.org/10.1016/J.IFACOL.2019.06.067
16. Zhang, C., Song, W., Cao, Z., Zhang, J., Tan, P.S., Chi, X.: Learning to dispatch for job shop scheduling via deep reinforcement learning. Adv. Neural. Inf. Process. Syst. **33**, 1621–1632 (2020)
17. Tang, J., Salonitis, K.: A deep reinforcement learning based scheduling policy for reconfigurable manufacturing systems. Procedia CIRP **103**, 1–7 (2021). https://doi.org/10.1016/J.PROCIR.2021.09.089
18. Evans, G.W., Mollaghasemi, M., Russell, E.C., Biles, W.E., Thompson, W.B.: In: Proceedings of the 1993 Winter Simulation Conference a tutorial for modeling with the witness visual interactive simulator (1993)
19. Mnih, V., et al.: Asynchronous methods for deep reinforcement learning. PMLR, pp. 1928–(1937). 11 Jun 2016. Accessed 13 Jul 2023. https://proceedings.mlr.press/v48/mniha16.html
20. Mnih, V., et al.: Human-level control through deep reinforcement learning. Nature **518**(7540), 529–533 (2015). https://doi.org/10.1038/nature14236

Parameter Identification in Manufacturing Systems Using Physics-Informed Neural Networks

Md Meraj Khalid and René Schenkendorf(✉)

Automation and Computer Sciences Department, Harz University of Applied Sciences, Wernigerode, Germany
rschenkendorf@hs-harz.de

Abstract. In this paper, we explore the application of Physics-Informed Neural Networks (PINNs) in parameter identification for continuum models of manufacturing systems. Although these models are invaluable for production planning at the factory level, the reliability of model-based decision-making strategies hinges significantly on accurate parameter estimation. We emphasize the distinct differences between PINNs and conventional parameter identification methods, particularly in terms of parameter sensitivities and uncertainty quantification. Our findings reveal that the PINN-based identification framework results in more significant parameter uncertainties. Consequently, this prompts us to discuss the implications for experimental designs, system identification, and the pivotal role of smart data.

Keywords: manufacturing systems · physics-informed neural network · partial differential equations · distributed parameter system · parameter sensitivities · uncertainty quantification

1 Introduction

Model-based methodologies have seen an increasing adoption in supply chain management. However, their efficacy is closely linked to the reliability of the data and manufacturing system models employed [7]. For instance, numerous manufacturing control strategies at the factory level resort to rudimentary fluid models or their more advanced counterparts. These models aim to emulate manufacturing systems' stochastic and non-linear dynamics [3,4,6]. To this end, accurate parameter identification is crucial for extracting meaningful insights regardless of the chosen model.

The literature predominantly describes standard identification methods for continuum models of manufacturing systems as a series of steps. Initially, the underlying partial differential equations (PDEs) are discretized. This is followed by a numerical solution of the resulting differential equation system, and it ends with parameter identification through local or global optimization routines. Furthermore, these estimated parameter values may be enhanced by uncertainty

A. Wagner et al. (Eds.): ESAIM 2023, LNME, pp. 51–60, 2024.
https://doi.org/10.1007/978-3-031-57496-2_6

quantification. In this context, local parameter sensitivity measures, such as the inverse of the Fisher Information Matrix and the Cramér-Rao inequality, are frequently utilized [13]. However, these methods, which include discretization, numerical solutions of differential equations, and optimization algorithms, are by no means free of challenges [1]. As a result, the standard parameter identification framework can inadvertently lead to approximation errors, thereby affecting the precision of model predictions and subsequent supply chain management and control strategies.

Physics-Informed Neural Networks (PINNs) introduce an alternative approach, offering simultaneous solutions of PDEs (i.e., forward problem) and model parameter identification (i.e., inverse problem) [11,16]. However, the impact of PINNs on system identification and model parameter estimates requires further research. For example, preliminary studies suggest differences between standard parameter identification and parameter estimation based on PINNs and scientific machine learning in terms of parameter uncertainties [14,15]. Therefore, this work contributes to an initial investigation into the effectiveness of using PINNs for solving continuum models and identifying parameters within manufacturing systems. As far as the authors are aware, this specific aspect has not yet been addressed in the existing literature.

The paper is organized as follows. In Sect. 2, information is provided on models for manufacturing systems, PINNs, and system identification. Section 3 discusses a case study using a continuum model for a single production line problem. Section 4 concludes the paper.

2 Methods

2.1 Manufacturing System Modeling

Research in supply chain management includes decision policies based on model-based and control concepts. Accurate modeling of manufacturing systems is crucial to predict system performance, optimize production processes, and ensure efficient resource allocation. Discrete event simulations (DES) have been used in the literature to study manufacturing processes, especially those with significant stochasticity and nonlinearity. However, while effective, DES can become exceedingly complex and challenging to analyze for larger systems[7,13].

To address this complexity issue and provide a more tractable representation of these systems, the "plug-flow" fluid model has been introduced. The fluid model approach assumes constant material flow throughout the manufacturing system, regardless of workload or internal process dynamics. Despite its usefulness in developing various decision policies, there is an increasing demand for more accurate fluid models that do not impose high computational costs for applications close to real-time [2,6]. In response to this demand, the continuum model combines the concepts of mass conservation and queueing theory's continuous queue length.

The continuum model uses a distinctive variable, termed the *stage*, to track the percentage of progress of each item in production. As an item enters the factory, this stage, represented by x, starts from 0 (indicating completion of 0%) and reaches 1 (signifying completion of 100%) as the item leaves the factory [3,10]. Although the stage variable x provides insight into the progression of the items, it does not necessarily represent their physical location within the factory [8,9]. However, each specific stage x can be aligned with a particular physical manufacturing process.

The continuum model assumes a product density (i.e., items per stage) $\rho(x,t)$ at a given stage x and a given time t. Thus, in mathematical terms, the continuum model is a partial differential equation, resulting in a distributed parameter system (DPS) problem, for which the specific form of the PDE, along with its boundary and initial conditions, will be detailed in the following subsection.

2.2 Distributed Parameter System

Distributed parameter systems (DPS) are mathematical models described by partial differential equations (PDEs) with an infinite-dimensional state space. Such models are crucial in various applications, including manufacturing systems. In the context of manufacturing, continuum models typically employ PDEs, with solutions given by $\rho(x,t)$, including model parameters θ, stage x, time t, a given domain called Ω, and $\partial\Omega$ for the boundary of Ω. The mathematical representation of the continuum model is given as [11]:

$$f\left(x, \frac{\partial \rho}{\partial x}, \frac{\partial^2 \rho}{\partial x^2}; \theta\right) = 0, \quad x \in \Omega, \tag{1}$$

$$\mathscr{B}(\rho, x) = 0 \text{ on } \partial\Omega, \tag{2}$$

where $f(\cdot)$ denotes the differential operator, and $\mathscr{B}(\cdot)$ represents the boundary operators.

Identifying and controlling DPS models is challenging. Often, DPS-based solutions are non-analytical, leading to the necessity for numerical approximation methods. Conventional techniques, such as the method of lines (MOL) or spectral methods, can be computationally intensive [12]. Moreover, selecting a suitable numerical method often depends on the PDE's particular form and the required precision of the solution.

By employing the MOL approach to Eq. (1), a system of ordinary differential equations (ODE) is derived. Subsequently, standard numerical routines can be used to solve the associated initial value problem. However, simulating the PDE model requires knowledge of all parameters, θ, such as kinetic coefficients or unknown disturbances. This need leads to the actual parameter identification challenge [13]. Some of these identification problems can be well-posed with unique solutions, while others might be ill-posed, requiring intensive efforts for accurate parameter estimation [1].

In contrast to the standard parameter identification framework that involves discretization, ODE solutions, and parameter estimation, PINNs offer a notable alternative. PINNs enable simultaneous PDE system solutions and parameter identification without the necessity of discretization or numerical ODE solvers

[16]. As a result, the PINN method combines essential aspects of system identification into one streamlined processing step, making it an effective and efficient approach in terms of implementation.

2.3 Physics-Informed Neural Networks

Introduced by Karniadakis et al. [11], the concept of Physics-Informed Neural Networks has found wide acceptance in scientific machine learning across various domains. PINNs are particularly useful for solving inverse or ill-posed problems when data is limited. This is mainly because PINNs incorporate known physical laws (described by differential equations) into their structure, allowing them to make accurate predictions even in data-sparse scenarios. In contrast, conventional numerical methods are generally more suited for well-posed standard problems [5].

Particularly in material science, where inverse problems (i.e., parameter identification) are common, PINNs have proven useful. These neural networks simplify the identification of geometric and material parameters by eliminating the need for specific algorithms or regular re-calibration of the computational mesh used for discretization [5].

To address a distributed parameter system (as given by Eqs. (1) and (2)), a deep neural network, denoted as $\rho(x, t; \lambda)$, is utilized. Here, λ represents the parameters of the neural network. This network is trained on available data to minimize the discrepancy between its predictions and the known physical laws defined by the PDEs. The combination of the initial and boundary conditions (Eq. (2)) with the PDE system (Eq. (1)) formulates the actual identification problem when using the PINN-based approach.

2.4 System Identification Based on PINNs

System identification is crucial in scientific simulations, and PINNs offer a unique approach. In detail, PINNs use a multi-objective loss function to predict outcomes and identify parameters. The multi-objective loss function, in turn, reflects different performance measures and often includes [5]:

- *Data discrepancy*: A measure of the difference between the neural network's predictions and the known reference simulations or sensor data. For continuum models, these data typically come from in-silico experiments, such as discrete event simulations.
- *Physical discrepancy*: In our context, this quantifies the deviation between the neural network's predictions and the established physical laws or constraints relevant to the manufacturing system. This is specifically about Eqs. (1) and (2).

Mathematically, the applied multi-objective loss function is formulated as [5]:

$$\mathcal{L}(\lambda, \theta) = \alpha_d \mathcal{L}^{\text{data}}(\lambda) + \mathcal{L}^{\text{pde}}(\lambda, \theta). \tag{3}$$

Herein, the *data discrepancy* loss term, $\mathcal{L}^{\text{data}}(\lambda)$, represents the deviation of in-silico experiment data, ρ^{data}, at discrete values along the stage axis, x_i, and specific time points, t_j, as L^2-norm error (i.e., least squares error):

$$\mathcal{L}^{\text{data}}(\lambda) = \sum_{i=1}^{m} \sum_{j=1}^{K} \| \rho^{data}(x_i, t_j) - \rho(x_i, t_j; \lambda) \|_2^2. \tag{4}$$

The *physical discrepancy* loss term $\mathcal{L}^{\text{pde}}(\lambda, \theta)$, in turn, reads as:

$$\begin{aligned} \mathcal{L}^{\text{pde}}(\lambda, \theta) = \alpha_f \sum_{j,n} \left\| f\left(\rho(x_j, t_n; \lambda), \frac{\partial \rho(x_j, t_n; \lambda)}{\partial x}, \frac{\partial^2 \rho(x_j, t_n; \lambda)}{\partial x^2}; \theta \right) \right\|_2^2 \\ + \alpha_b \sum_{j,n} \| \mathscr{B}(\rho(x_j, t_n; \lambda); \theta) \|_2^2. \end{aligned} \tag{5}$$

In Eqs. (3) and (5), α_d, α_f, and α_b are weight factors of the corresponding loss term components, and $j \in \mathcal{T}_x$ and $n \in \mathcal{T}_t$ indicate the collocation points for the stage and time dimension. The prediction of the neural network, $\rho(x_j, t_n; \lambda)$, should align with the equation system and the in-silico data.

The efficacy of PINNs is based on the neural network's architecture, the data quality, and the loss-term weighting factors. In subsequent implementations, the default values of the weight factors α_d, α_f, and α_b are 1, i.e., the related loss-term contributions are considered equally important. A comprehensive overview of the PINN-based system identification process is presented in Fig. 1.

3 Case Study

In our investigation of the application of PINNs in manufacturing systems, we focus on a production line as our case study. This particular production line was chosen because it serves as a basic unit when analyzing a more comprehensive manufacturing system at the factory level [6,8]. Our chosen production line

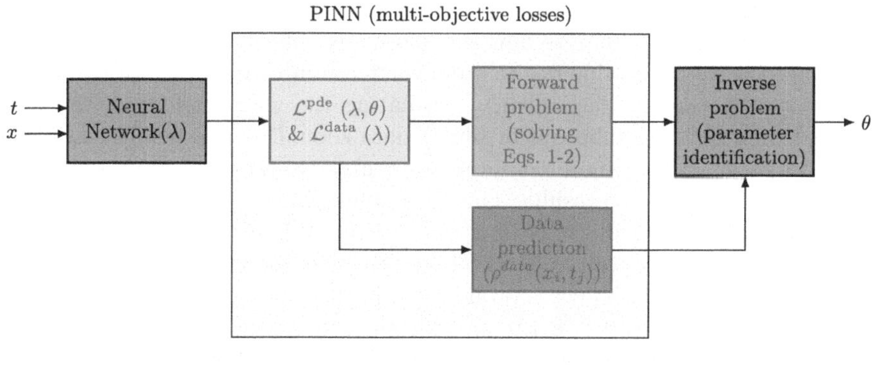

Fig. 1. PINN-based parameter identification scheme.

comprises a series of interconnected workstations. By adopting the continuum model approach, the behavior of the production line is dictated by a single PDE. We map each workstation in this production line to a stage coordinate value x, which ranges between 0 and 1. Specifically, $x = 0$ marks the inlet, and $x = 1$ signifies the outlet of the production line [2]. There are certain assumptions about this model: (i) items within the production line are conserved, (ii) the product density $\rho(x,t)$ and the flux $v(\rho(x,t))$ are non-negative [9].

The dynamic behavior of the production line is characterized by the following equations [9]:

$$\frac{\partial \rho(x,t)}{\partial t} = -\frac{\partial v(\rho(x,t))}{\partial x}, \tag{6}$$

representing the rate of change of product density over time, and

$$v(\rho(x,t)) = \frac{\mu \rho(x,t)}{M + \rho(x,t)}, \tag{7}$$

with the flux as a function of the product density. In these, μ is the average processing rate, and M is the number of workstations.

For the default *ramp-up* scenario, we chose $\mu = 2.0$, $M = 10$, $\rho(t,0) = 10$, $\rho(0,x) = 0$, and discretized the PDE with the MOL approach assuming $\Delta x = 0.05$. Figure 2 visualizes the production density $\rho(t,x)$ at time values $t \in \{1.5, 3.0, 4.5\}$. In this scenario, the first product does not exit the production line ($x = 1$) before 4.5 time units.

Next, a multilayer perceptron (MLP) neural network is used for the PINN framework. The MLP was chosen for its widespread acceptance in similar applications and its straightforward implementation, which balances complexity and capability. It consists of two hidden layers, each containing 16 neurons. The activation function for these layers is the sigmoid function, and the overall implementation is done in the Julia programming language, where the limited-memory BFGS (L-BFGS) optimizer was used to train the MLP with an iteration limit of 2500 and 1e-6 for the convergence threshold of the objective function.

The predicted production density $\rho(x,t)$ based solely on the PINN approach is shown in Fig. 3a. As the results are consistent with those obtained through the MOL, there are no noticeable differences to highlight. However, in Fig. 3b, the significance of the mean processing rate, denoted by μ, is illustrated. Here, we reduced μ to 0.5 to analyze how this parameter affects $\rho(t,x)$. As expected, the reduced parameter value resulted in a decrease in throughput. In turn, with the observed parameter dependency, estimating the value of μ from the data obtained from DEM simulations or sensor data should be practically feasible.

Figure 4a provides a more comprehensive understanding of the parameter's influence on the predicted results. This figure offers insight into the local parameter sensitivity, specifically regarding $d\rho(t,x)/d\mu$. When there are fluctuations in the mean processing rate, denoted by μ, they considerably affect the model predictions, $\rho(t,x)$. This effect is pronounced in the region where the stage coordinate, x, extends from 0.4 to 1.0, and the time, t, extends from 0 to 15 units. Such patterns suggest favorable parameter sensitivity in these regions. According to the Cramér-Rao inequality, areas of increased parameter sensitivity will

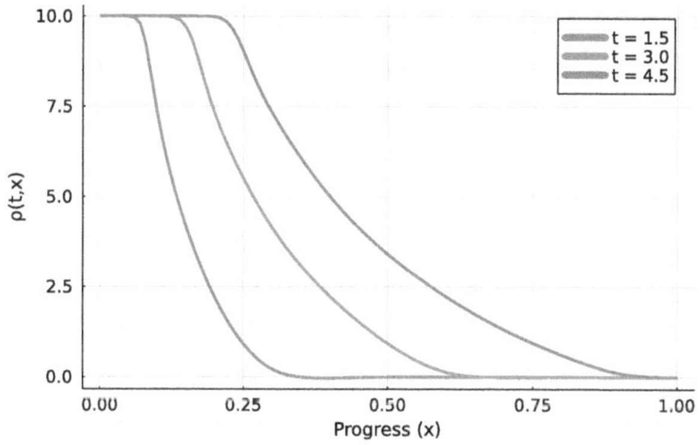

Fig. 2. Product density $\rho(t, x)$ for three different time values t for a ramp-up scenario.

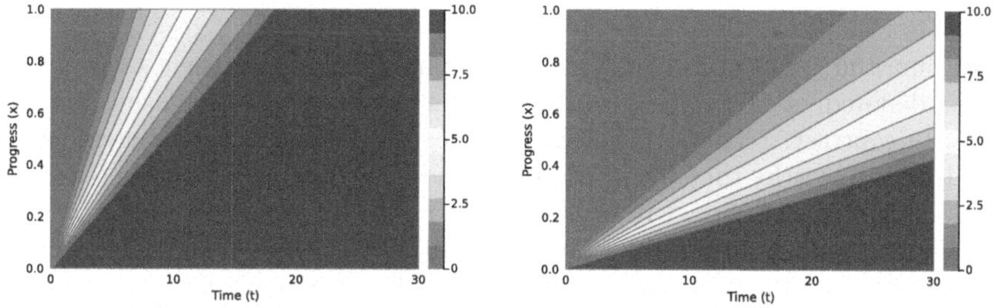

(a) Default rate ($\mu = 2.0$).

(b) Reduced rate ($\mu = 0.5$).

Fig. 3. Product density $\rho(t, x)$ as predicted by the Physics-Informed Neural Network.

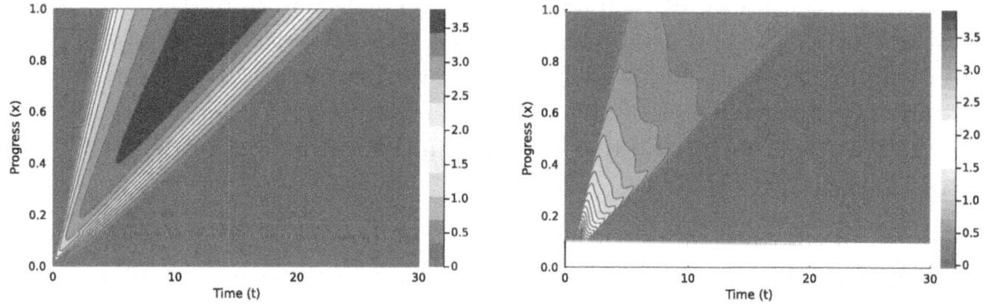

(a) Method of Lines.

(b) Physics-Informed Neural Network.

Fig. 4. Parameter sensitivity of the mean processing rate μ for the standard MOL-based and PINN-based framework.

likely result in more accurate estimations of μ. This holds true, at least for standard parameterization methods such as MOL, where numerical solutions of the differential equation precede the parameter identification step.

Next, we use different sampling scenarios to practically estimate the mean processing rate, μ.

Table 1. Parameter uncertainties of the mean processing rate parameter μ.

Time sampling scenarios	Standard deviation (MOL)	Standard deviation (PINN)
Scenario A ($\alpha_d = 1$)	0.0024	0.0709
Scenario B ($\alpha_d = 1$)	0.0018	0.0517
Scenario A ($\alpha_d = 5$)	-	0.0792
Scenario B ($\alpha_d = 5$)	-	0.1137
Scenario A ($\alpha_d = 10$)	-	0.1206
Scenario B ($\alpha_d = 10$)	-	0.1223

In our study, we explored two scenarios characterized by their sample time points, denoted as t_i. In Scenario A, t_i spans from 0 to 30, increasing in steps of 0.5. In contrast, Scenario B covers t_i values from 0 to 15, with increments of 0.25. It is noteworthy that the discretization for the stage coordinate x remained consistent in all scenarios, with x_i values set as $\{0, 0.1, \ldots, 1.0\}$.

Table 1 provides an insight into the variability of the estimated parameter μ, symbolized by its standard deviation, σ_μ. The results are influenced by the choice of reference points $\{x_i, t_i\}$ and the assumed "measurement noise" with a variance of 0.01. This noise can originate either from stochastic DEM simulations or direct sensor data. Predictably, Scenario B exhibited reduced parameter uncertainties. This observation aligns with the sensitivity patterns highlighted in Fig. 4a. It suggests that a higher density of reference points $\{x_i, t_i\}$ within regions of increased parameter sensitivity results in more precise parameter estimations. An essential point of reference: the second column of Table 1 presents the variability in predicted parameters when following a traditional approach, that is, employing MOL followed by a separate parameter estimation step. To determine the uncertainties of the resulting parameters for scenarios A and B via the PINN approach, Monte Carlo simulations were performed with 1000 repetitions each.

4 Discussion

When comparing the parameter uncertainties based on PINN in Table 1, it is clear that they are one order of magnitude worse than the standard parameter identification approach.

A detailed inspection of Fig. 4b provides insight into the root cause of this divergence. The PINN framework uses the loss term $\mathcal{L}^{\text{pde}}(\lambda, \theta)$ (Eq. (5)) to obtain parameter estimates. The parameter sensitivity, in turn, related to the loss term, is shown in Fig. 4b. It highlights the opposite behavior compared to the standard parameter identification approach. The origin of the x and t axes has crucial sensitivity values. Please note that in Fig. 4b, the sensitivities in the range of 0 to 0.1 for the x dimension were excluded to maintain a consistent sensitivity scale. Furthermore, lower ρ values are typical at the origin, and thus the assumed additive noise with a variance of 0.01 yields a poor signal-to-noise ratio, complicating the precise determination of the model parameter.

Consequently, when the PINN approach is used, it is important to note that it removes the requirement of discretization and numerical solving of the differential equation. Nevertheless, it requires estimating all neural network parameters λ. In our case study, this results in a total of 338 unknowns. Therefore, the information content of the reference data is distributed and used differently. This implies that the estimation of μ and its associated uncertainties through PINN is significantly influenced by other factors affecting the neural network, such as the number of hidden layers, neurons, activation functions, and weights of the multi-objective loss term. For example, as shown in Table 1, the weight factor α_d can differ and is part of the set $\alpha_d \in \{1, 5, 10\}$.

As the weighting factor increases, the uncertainties also increase. This is probably because the PINN method aims to fit noisy data, which may affect the PDE component (Eq. (5)) in the loss term (Eq. (3)). Thus, the influence of the actual continuum model is reduced, including the mean processing rate parameter μ. If Eq. (5) is used with higher weighting factors in the loss function, it may result in overfitting. This happens when data fitting is given more importance than capturing the model equations of the underlying continuum model. Therefore, the weight factors and the general specifications of the neural network should be optimally selected.

While PINNs promise to simplify the parameter determination process, how they compute loss terms can influence parameter sensitivity and, subsequently, the associated parameter uncertainty. This change in sensitivity and uncertainty means that traditional methods for selecting reliable reference data and achieving accurate parameter estimates might not be as effective.

5 Conclusions

In conclusion, this study represents a pioneering effort to apply Physics-Informed Neural Networks (PINNs) in identifying parameters within continuum models of manufacturing systems. Although our findings demonstrate that PINNs can be employed to solve continuum models in manufacturing systems, they did not show significant improvement in parameter accuracy for our specific case study. Therefore, future research will be directed toward formulating specific strategies for data selection and weight factor adjustments in PINNs, to improve parameter estimates and reduce uncertainties.

References

1. Abt, V., et al.: Model-based tools for optimal experiments in bioprocess engineering. Curr. Opin. Chem. Eng. **22**, 244–252 (2018). https://doi.org/10.1016/j.coche.2018.11.007
2. D'Apice, C., Kogut, P.I.: On optimal control problem for conservation law modelling one class of highly re-entrant production systems. In: AIP Conference Proceedings. Author(s) (2017). https://doi.org/10.1063/1.4992738
3. Diagne, M., Karafyllis, I.: Event-triggered boundary control of a continuum model of highly re-entrant manufacturing systems. Automatica **134**, 109902 (2021). https://doi.org/10.1016/j.automatica.2021.109902
4. Göttlich, S., Herty, M., Klar, A.: Modelling and optimization of supply chains on complex networks. Commun. Math. Sci. **4**(2), 315–330 (2006). https://doi.org/10.4310/cms.2006.v4.n2.a3, https://doi.org/10.4310%2Fcms.2006.v4.n2.a3
5. Karniadakis, G.E., Kevrekidis, I.G., Lu, L., Perdikaris, P., Wang, S., Yang, L.: Physics-informed machine learning. Nat. Rev. Phys. **3**(6), 422–440 (2021). https://doi.org/10.1038/s42254-021-00314-5
6. Marca, M.L., Armbruster, D., Herty, M., Ringhofer, C.: Control of continuum models of production systems. IEEE Trans. Autom. Control **55**(11), 2511–2526 (2010). https://doi.org/10.1109/tac.2010.2046925
7. Mönch, L., Fowler, J.W., Mason, S.J.: Production Planning and Control for Semiconductor Wafer Fabrication Facilities. Springer, New York (2013). https://doi.org/10.1007/978-1-4614-4472-5
8. Othman, K.A., Meurer, T.: Demand tracking control in manufacturing systems. IFAC-PapersOnLine **53**(2), 11219–11224 (2020). https://doi.org/10.1016/j.ifacol.2020.12.334
9. Othman, K.A., Meurer, T.: Optimal boundary control for the backlog problem in production systems. IFAC-PapersOnLine **55**(20), 511–516 (2022). https://doi.org/10.1016/j.ifacol.2022.09.146
10. Pihnastyi, O., Yemelianova, D., Lysytsia, D.: Using PDE model and system dynamics model for describing multi-operation production lines. Eastern-European J. Enterp. Technol. **4**(4 (106)), 54–60 (2020). https://doi.org/10.15587/1729-4061.2020.210750
11. Raissi, M., Perdikaris, P., Karniadakis, G.: Physics-informed neural networks: a deep learning framework for solving forward and inverse problems involving nonlinear partial differential equations. J. Comput. Phys. **378**, 686–707 (2019). https://doi.org/10.1016/j.jcp.2018.10.045
12. Schiesser, W.E., Griffiths, G.W.: Compendium of Partial Differential Equation Models. Cambridge University Press, Cambridge (2009)
13. Schwartz, J.D., Rivera, D.E.: A system identification approach to PDE modeling of a semiconductor manufacturing process. IFAC Proc. Volumes **42**(10), 964–969 (2009). https://doi.org/10.3182/20090706-3-fr-2004.00160
14. Selvarajan, S., Tappe, A.A., Heiduk, C., Scholl, S., Schenkendorf, R.: Process model inversion in the data-driven engineering context for improved parameter sensitivities. Processes **10**(9), 1764 (2022). https://doi.org/10.3390/pr10091764
15. Tappe, A.A., Schulze, M., Schenkendorf, R.: Neural ODEs and differential flatness for total least squares parameter estimation. IFAC-PapersOnLine **55**(20), 421–426 (2022). https://doi.org/10.1016/j.ifacol.2022.09.131
16. Yazdani, A., Lu, L., Raissi, M., Karniadakis, G.E.: Systems biology informed deep learning for inferring parameters and hidden dynamics (2019). https://doi.org/10.1101/865063

Detection of Rare Fault Cases for Mobile Robot Applications

Benjamin Blumhofer(✉), Jonas Weigand, Leonhard Kunz, Pascal Rübel, and Achim Wagner

German Research Center for Artificial Intelligence, Kaiserslautern, Germany
benjamin.blumhofer@dfki.de

Abstract. Modern production and transport systems consist of many individual mechanical components, sensors, and actuators. Despite the complexity, critical errors must be diagnosed quickly to avoid the propagation of faults and the associated safety risks and production downtime. The problems are incomplete descriptions of the systems and disturbed sensor signals, which makes it difficult to use model-based methods directly. In this article, a method is presented that is based on the description of dynamical systems using Runge-Kutta Neural Networks (RKNN) as nominal models. RKNNs have a good ability to predict dynamic processes and can be trained with comparatively little data. Deviations from the real process behavior to the nominal behavior lead to generic fault symptoms classified by a second neural network. The fault cases are purely modeled with the help of the process data without providing special knowledge of the internal system behavior. The procedure was implemented and tested on a mobile robot for factory-internal transport processes. With little training data, a binary classification for elementary fault detection. The method can easily be adapted to other cases of dynamical systems.

Keyword: fault detection · mobile robotics · neural networks · ordinary differential equations

1 Introduction

The introduction highlights the increasing complexity in the design and control of industrial production facilities, driven by the demand for customized products and the need for flexibility and quality assurance. Modular production and transport processes are proposed as an approach to address this complexity, with manufacturing cells and Automated Guided Vehicle (AGV) playing crucial roles. Model-based engineering approaches using mathematical models or digital twins are discussed as tools for controlling and optimizing production processes, as well as predicting and diagnosing machine faults. However, the high cost and limited availability of control experts make data-based methods appealing, particularly using Neural Networks (NNs) to create surrogate models. While NNs can achieve good performance with ample training data, there are challenges related to the availability and distribution of input data, predictability from noisy or limited data, instability in certain types of neural networks, and difficulties in interpreting and validating surrogate models, especially in safety-critical applications.

A. Wagner et al. (Eds.): ESAIM 2023, LNME, pp. 61–70, 2024.
https://doi.org/10.1007/978-3-031-57496-2_7

In this article, a surrogate model approach is provided for Fault Detection (FD). The focus is on Runge-Kutta Neural Networks (RKNNs) since they can be augmented with stability criteria, and they require relatively little training data compared to the prediction accuracy [23]. The explainability of the models can be achieved by separating the linear from the non-linear fractions and by comparing them to linear reference models. To demonstrate the feasibility of RKNNs within a mobile robot FD scenario, we apply an RKNN-based surrogate model on the robot motion control level. This article is divided into three major parts, Sect. 2 and Sect. 3 discuss the theory of the approach, Sect. 4 explains the methodology of this work, and Sect. 6 shows the implementation of our experiment.

2 Related Work

Combining physical and statistical models for fault detection and identification in dynamical systems is an evident approach, as analytical models have already been developed for many systems. These deterministic models often incorporate extensive knowledge about the underlying physical dynamics. Nevertheless, they usually lack expressiveness to represent non-linear effects, like noise, disturbances, and errors, as well as complex physical effects, i.e., friction. Statistical models can cover these effects. Zak and Park (2001) [24] used the term gray-box models here because these approaches use a deterministic white box model in combination with a statistical black box model. Mackey et al. (2001) [18] in-corporate this hybrid approach in an end-to-end framework for real-time fault detection and characterization called Beacon-based Exception Analysis for Multimissions (BEAM). The BEAM architecture is further optimized and extended and integrated into a health management system for autonomous probes and vehicles for extraplanetary missions [20].

The combination of Ordinary Differential Equation (ODE) and NNs as pro-posed by Chen et al. (2018) [2] has also been investigated in the context of Fault Detection and Isolation (FDI) methods [4, 13]. These approaches have been applied in different domains like engine or exhaust faults [14, 15] or Unmanned Areal Vehicle (UAV) dynamics [5]. The authors hereby use first order ODE-solvers.

RKNN methods employ a fourth-order solver to gain higher prediction accuracy [22]. Despite a more accurate system identification with RKNNs, which is essential for reliable residual calculation, to the authors' knowledge, there has not been any approach to applying these methods to FD.

3 Theoretical Foundations

3.1 Fault Detection

Due to the increasing complexity and scale of industrial systems, there is less and less tolerance for performance and productivity degradation and safety risks. To increase the reliability and safety of industrial systems, it is essential to detect faults as early as possible [8]. The last few decades have seen the development of many different approaches to this challenge, with various methodical segmentations of the underlying problem [7].

This schema of [12] describes fault detection through the detection of changes between an observed normal or nominal system behavior and the currently observed systems features. This approach does not explicitly model the process as a change of the internal state due to actuator inputs. Isermann (1997) [11] derived requirements for the nominal model from which two are interested to this work:

I Early detection of small faults with abrupt or incipient time behavior,
II Diagnosis of faults in the actuator, process components or sensors.

3.2 Nominal Models

Nominal models estimate the nominal state of a process based on previously observed behavior and additional sensor data. This approach applies the time-dimension as source of information, compared to other approaches using cross-feature information. Time-based nominal models use self-supervised learning, i.e. a measured past state $\hat{x}_{n-1} \in \mathbb{R}^{N_x}$ with $N_x \in N$ observable states and the discrete time step $n \in N$, is the source of information to predict the nominal state x_n. . Note that for Fault Detection and Diagnosis (FDD) a time-prediction is not obligatory. Comparing a current measurement $\hat{x}_n$ with an expected, nominal state x_n is sufficient. Cross-feature models estimate each component $x_{i,n}$ of the expected, nominal state $x_n = \left[x_{1,n}, x_{2,n}, \ldots, x_{N_x,n}\right]$ using sensory data of the other state components, e.g. estimation of $x_{2,n}$ uses the information from $\left[\hat{x}_{1,n}, \hat{x}_{2,n}, \ldots, \hat{x}_{N_x,n}\right]$. Both, time, and feature based nominal models, do not re-quire labeled training data - and thus significantly reduce implementation effort and scalability of the method. Regarding FDD, cross-feature models are in general more appropriate in detecting drifting anomalies, and prediction-models are suitable for fast, sudden anomalies. However, the performance of cross-feature versus prediction-based models depends on the concrete configuration and application, too.

In the field of time series prediction with NN, discrete-time models can be distinguished from continuous-time models. Popular examples of discrete-time models are Recurrent Neural Network (RNN), Gated Recurrent Units (GRU) or Long Short-Term Memory (LSTM) [3, 10] and examples of continuous-time NN are RKNN, Liquid Time Constant Neural Network (LTC), and Neural Ordinary Differential Equation (NODE) [2, 9, 22]. A comprehensive survey on continuous time NN is given in [1, 6, 16]. Discrete-time models represent the observed data, whereas continuous-time NN capture the underlying, continuous process. Continuous-time models are data efficient, can naturally handle irregularly sampled data [17], and can simply be enhanced by physical expert knowledge in terms of differential equations. At its core, continuous-time models combine an ODE solver with a NN - in case of RKNN, the Runge-Kutta ODE solver is applied.

The RKNN gets the state x(t) and additional features u(t) $\in \mathbb{R}^{N}$U as in-put and computes the derivative of the hidden state x˙(t) $\in \mathbb{R}^{N}$X as output. In addition, predictions y(t) $\in \mathbb{R}^{N}$Y are made based on the state x(t) and input u(t),

$$x'(t) = f_s(x(t), u(t)) = A_{NN}x(t) + B_{NN}u(t) + f_{NN}(x(t), u(t)), \tag{1a}$$

$$y(t) = h_s(x(t), u(t)) = C_{NN}x(t) + D_{NN}u(t) + h_{NN}(x(t), u(t)). \tag{1b}$$

For classification tasks, $y(t)$ contains the output distribution, so $N_Y \in N$ is equal to the number of target classes. For regression tasks, the number of output channels N_Y corresponds to the number of independent target variables to predict.

Note that (1) is enhanced by a linear core defined as state-space model matrices $A_{NN} \in \mathbb{R}^{N_x \times N_x}$, $B_{NN} \in \mathbb{R}^{N_x \times N_u}$, $C_{NN} \in \mathbb{R}^{N_y \times N_x}$, and $D_{NN} \in \mathbb{R}^{N_y \times N_u}$.

This formulation is analog to the state-space model and enables direct incorporation of physical knowledge, even on a network parameter level.

To estimate (1) on discrete data, the 4-stage Runge-Kutta algorithm is applied as ODE solver, as explained in [23].

The initial state n_0 is estimated using a Deep Encoder Network (DEN) [1] using the information of a past horizon of input and output variables. It is a very flexible approach without losing information. However, DENs introduce an additional model, which increases computational effort and memory load and requires sufficiently rich data to be trained.

4 Methodology

We developed a process based on the requirements of FD approaches. The process consists of four major steps, including the data acquisition of sensor data, the model of the system's nominal behavior, the generation of the residual, and fault detection which is shown in Fig. 1. To meet requirement, I to detect early and small faults in the time behavior, there is a need for a precise nominal model to detect suspicious changes in time series data as early as possible. We use a RKNN to meet this requirement to predict the fault-free system. Requirement II demands the diagnosis of faults in the entire system, which is not the part of this article.

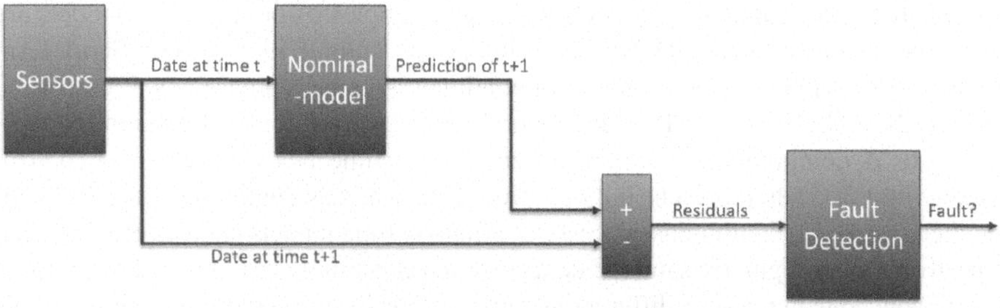

Fig. 1. Schematic drawing of the FD approach

One of the key elements of this FD process is the nominal model, which is based on an RKNN. A small set of fault-free measurements are recorded for training. The major advantages of RKNNs is data efficiency, the ability to generalize and operate in situations not seen during the training. This significantly reduces manual data collection efforts. Furthermore, RKNN are partially explainable due to the similarity of the linear part to state space models. Those advantages match the requirements of nominal models. The need for only a few datasets is interesting because this usually takes a lot of resources to build. In general, RKNN are suitable for this challenge.

The network provides non-linear black box identification capabilities, which is essential [23]. In the scenario of this work, the precision of long-time prediction is not relevant but should be taken into account depending on the use case of the fault detection process. In our implementation, one-step prediction is sufficient and more accurate.

The prediction calculated in the nominal model is used to calculate the residuals between the predicted and measured state to detect faults. There are multiple approaches to classify faults based on the given residuals. As explained in Sect. 3, [12] suggests using NNs for classification. This work follows this suggestion and uses RKNN for fault classification. As this article focuses on pure fault detection the problem ca be described as a binary classification scenario.

5 Mobile Robot

Experiments are carried out on the Festo Didactics Robotino 4 shown in Fig. 2. We exploit on the internal sensors of the robot, especially those signals related to the locomotion system. In addition to the internal sensors, we utilize the acceleration measurements from Inertial Measurement Unit (IMU) within an Intel Realsense D435i Camera mounted on the robot. The robot is equipped with an omnidrive system consisting of three mecanum wheels.

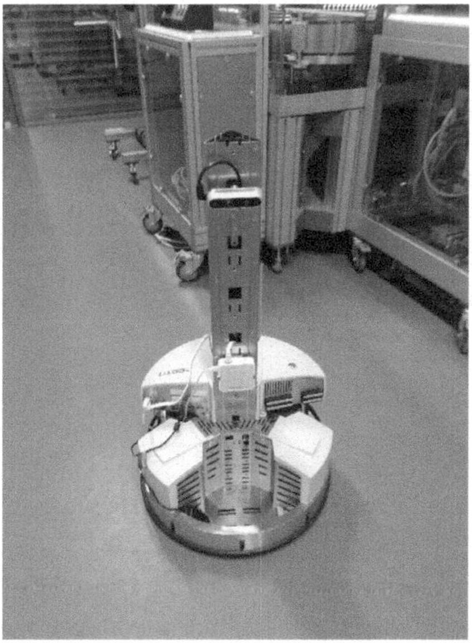

Fig. 2. Festo Robotino

To interact with the Robotino this work utilizes Robot Operating System (ROS) [21]. To run the fault detection algorithms on the Robot it is equipped with a Nvidia Jetson Nano which has one core assigned to the fault detection application.

6 Experiment

This section presents the results of the fault detection algorithms and the specifics of the conducted experiment. The inference were run on an Nvidia Jetson nano [19] which was able the run the pipeline within about 50 ms on one CPU core.

6.1 Setup of the Experiment

To test the proposed fault detection and diagnosis approach, we test three different scenarios. At first, we create a scenario where the camera including the IMU sensor is disconnected during the data acquisition, which results in not receiving sensor readings anymore. The second scenario contains wrong measurements of the IMU created by shaking and turning the camera. For the third scenario, the movement base of the Robotino is blocked. While the first two scenarios show faulty signals on a sensor, the third scenario contains valid signals in all sensors. The sensors give contradicting information that has to be detected. In Fig. 3, for each scenario, a human-readable deviation from the expected behavior is shown. The data used for training is pre-recorded and handled with a sample rate of 50 Hz.

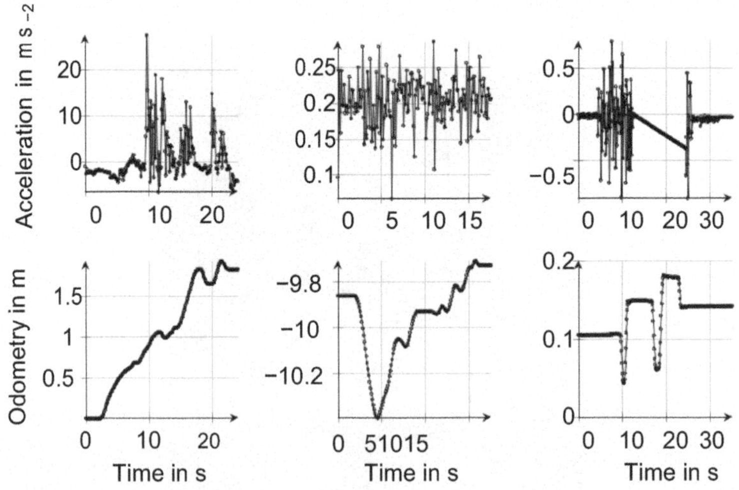

Fig. 3. Szenarios for FD. Left: Camera IMU shake. Middle: Jacked up robot. Right: Camera connection loss.

We acquire only small datasets for training to validate the advantage of RKNN regarding data efficiency. Therefore, we use only one dataset of 35 s navigating through a factory environment. The entire dataset of faults consists of about 110 s of measurements. The developed pipeline is segmented into three steps. First, the nominal state is predicted. Based on these predictions and the measurements of mobile robot, residuals are estimated and fed into a second Artificial Intelligence (AI) for classification. After the classification, steps of post-processing are performed.

6.2 Nominal Model of the Locomotion System

This section explains the network setup, training, and validation process. We apply 16 states in the RKNN, which are 3 encoder positions, 3 odometry positions and 4 orientations in quaternions, 3 linear accelerations and 3 angular velocities measured by the IMU. The 3 motor currents are used as exogenous input. The nonlinear networks f_{NN} and h_{NN} are Multi Layer Perceptrons (MLPs) with 2 hidden layers and LeakyReLU activation functions each. Including the linear state space model, A_S, B_S, C_S and DS, this results in 11,286 trainable parameters. Training on an Intel i7 – 1280 h and an Nvidia A2000 graphics card required less than 30 min. A test dataset is presented to the network in the Fig. 4. It is observable that the model cannot replicate the behavior perfectly, but it is sufficient to continue with fault detection.

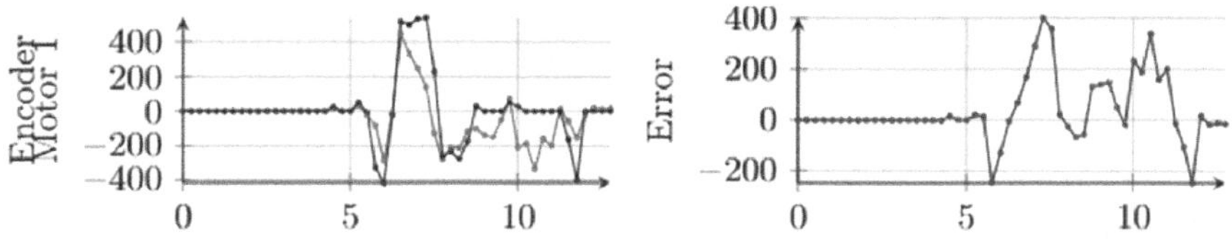

Fig. 4. Prediction by the nominal model and measurements of (unitless) motor encoder signals in the test dataset. Prediction (red), measurement (black) and prediction error (blue).

6.3 Fault Detection

In this section, we show the results of binary classification used for fault detection. To generate the input data for those algorithms, the residuals of the prediction and measurement of each timestep are calculated. In Fig. 5, the prediction and measurements of the fault caused by shaking the IMU are shown. It is visible that the odometry signal is highly influenced by the signal of the IMU, and the deviation between prediction and measurements start to increase when the camera is unmounted and starts shaking at around 9 s.

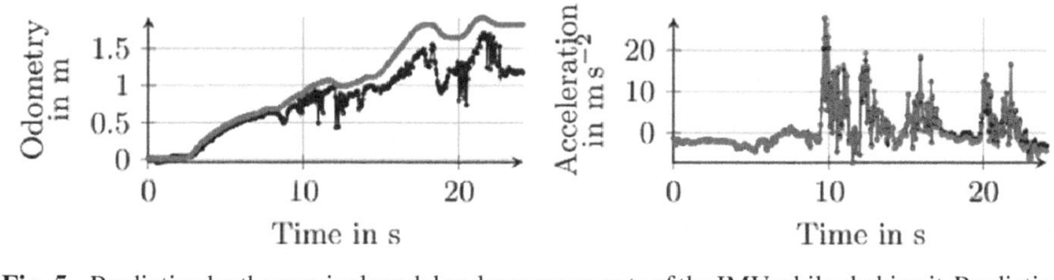

Fig. 5. Prediction by the nominal model and measurements of the IMU while shaking it. Prediction (red) and measurement (black).

For binary fault classification, we again use an RKNN structure. As shown in Fig. 4, the RKNN can classify faultily and fault-free datasets based on the residuals as input data. The binary classification uses the calculated residuals from Subsect. 6.2. The model consists of 4,503 trainable parameters outputting a binary state. The results in Fig. 6 show that the network can detect the faults without any significant falsely detected states. The only false classified states are at the beginning of a fault where the network detects the fault one timestep later than it occurred.

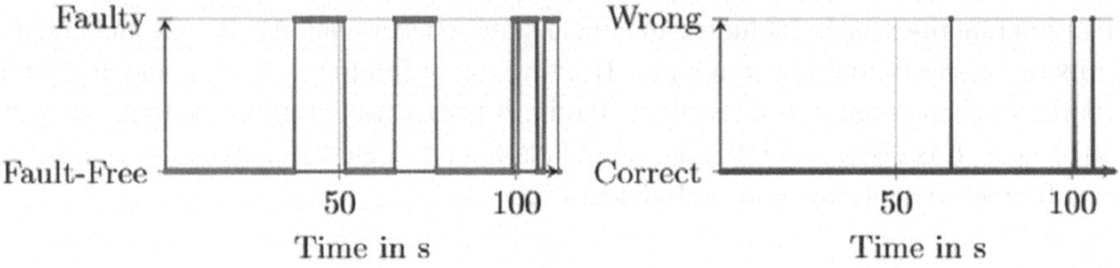

Fig. 6: Binary classification of all three faulty datasets. Left: Predicted (red) and ground-truth (black) fault state. Right: Correctly or wrong identified state (blue).

7 Discussion

This section discusses the results of the fault detection approach we implemented in this work, based on the proposed requirements in Sect. 3. The first requirement presented by [12] is related to detecting small faults with abrupt or incipient time behavior. It was tested by shaking and disconnecting the camera module (abrupt time behavior) and driving while the robot is jacked up (incipient behavior). Both faults were detected in our classification with negligible delay. In general, the results of this work show that RKNN are a suitable approach for fault detection. Compared to other data-driven techniques, the advantages in explainability, the demand for little data, and fast inference times are relevant in this application. In the use case chosen for this work, the long-time prediction quality of RKNN is not relevant, but it can become an advantage when detecting faults in other applications. The main research question of this paper, if RKNN are suitable for fault detection, is answered positively. A future research direction would be an increase in the complexity of the scenarios and a comparison with other methods, including and beyond neural networks. Additionally, the research could be expanded to evaluate the capabilites of RKNNs in the identification of faults. Beyond the identification of faults it would be interesting to increase prediction time for predictive maintenance use cases similar to [25] using RKNNs.

8 Conclusion

In conclusion, this paper has presented a novel approach for fault detection using RKNNs. The proposed method involves creating a nominal model from the RKNN and then calculating the residuals by subtracting the measurements from the nominal state estimation. The obtained residuals are then used to classify the faults. The results of this work demonstrate that RKNNs are a promising and effective approach for fault detection. This method offers robustness and computational efficiency compared to traditional fault detection approaches while being partially explainable.

References

1. Beintema, G., Toth, R., Schoukens, M.: Nonlinear state-space identification using deep encoder networks. In: Learning for Dynamics and Control, pp. 241–250. PMLR (2021)
2. Chen, R.T., Rubanova, Y., Bettencourt, J., Duvenaud, D.K.: Neural ordinary differential equations. In: Advances in Neural Information Processing Systems, vol. 31 (2018)
3. Chung, J., Gulcehre, C., Cho, K., Bengio, Y.: Empirical evaluation of gated recur-rent neural networks on sequence modeling (2014). arXiv preprint arXiv:1412.3555
4. Enciso-Salas, L., P´erez-Zu˜niga, G., Sotomayor-Moriano, J.: Fault diagnosis via neural ordinary differential equations. Appl. Sci. **11**(9), 3776 (2021)
5. Enciso-Salas, L., P´erez-Zu˜niga, G., Sotomayor-Moriano, J.: Fault detection and isolation for UAVs using neural ordinary differential equations. IFAC-PapersOnLine **55**(6), 643–648 (2022)
6. Forgione, M., Piga, D.: Continuous-time system identification with neural net works: model structures and fitting criteria. Eur. J. Control. **59**, 69–81 (2021)
7. Gao, Z., Cecati, C., Ding, S.: A survey of fault diagnosis and fault-tolerant techniques—part ii: Fault diagnosis with knowledge-based and hybrid/active approaches. IEEE Trans. Ind. Electron. **62**, 1–1 (2015). https://doi.org/10.1109/TIE.2015.2419013
8. Gao, Z., Cecati, C., Ding, S.X.: A survey of fault diagnosis and fault-tolerant techniques—part i: fault diagnosis with model-based and signal-based approaches. IEEE Trans. Industr. Electron. **62**(6), 3757–3767 (2015)
9. Hasani, R., Lechner, M., Amini, A., Rus, D., Grosu, R.: Liquid time-constant networks. In: Proceedings of the AAAI Conference on Artificial Intelligence, vol. 35, pp. 7657–7666 (2021)
10. Hochreiter, S., Schmidhuber, J.: Long short-term memory. Neural Comput. **9**(8), 1735–1780 (1997)
11. Isermann, R.: Supervision, fault-detection and fault-diagnosis methods—an introduction. Control. Eng. Pract. **5**(5), 639–652 (1997)
12. Isermann, R.: Fault-Diagnosis Applications - Model-Based Condition Monitoring: Actuators, Drives, Machinery, Plants, Sensors, and Fault-tolerant Systems. Springer Science and Business Media (2011). https://doi.org/10.1007/978-3-642-12767-0
13. Jung, D.: Residual generation using physically-based grey-box recurrent neural networks for engine fault diagnosis (2020). arXiv preprint arXiv:2008.04644
14. Jung, D.: Automated design of grey-box recurrent neural networks for fault diagnosis using structural models and causal information. In: Learning for Dynamics and Control Conference, pp. 8–20. PMLR (2022)
15. Jung, D., Kleman, B., Lindgren, H., Warnquist, H.: Fault diagnosis of exhaust gas treatment system combining physical insights and neural networks. IFAC-PapersOnLine **55**(24), 97–102 (2022)
16. Kidger, P.: On neural differential equations (2022). arXiv preprint arXiv:2202.02435
17. Kidger, P., Morrill, J., Foster, J., Lyons, T.: Neural controlled differential equations for irregular time series. Adv. Neural. Inf. Process. Syst. **33**, 6696–6707 (2020)
18. Mackey, R., James, M., Park, H., Zak, M.: BEAM: technology for autonomous self-analysis. In: 2001 IEEE Aerospace Conference Proceedings (Cat. No. 01TH8542), vol. 6, pp. 2989–3001. IEEE (2001)
19. Nvidia: Jetson nano developer kit user guide (2019). https://developer.download.nvidia.com/embedded/L4T/r32-3-1_Release_v1.0/Jetson_Nano_Developer_Kit_User_Guide.pdf. Accessed 15 Jan 2023
20. Park, H., Mackey, R., James, M., Zak, M., Baroth, E.: BEAM: technology for autonomous vehicle health monitoring (2002)

21. Stanford Artificial Intelligence Laboratory et al.: Robotic operating system. https://www.ros.org
22. Wang, Y.J., Lin, C.T.: Runge-kutta neural network for identification of dynamical systems in high accuracy. IEEE Trans. Neural Networks **9**(2), 294–307 (1998)
23. Weigand, J., Deflorian, M., Ruskowski, M.: Input-to-state stability for system identification with continuous-time Runge–Kutta neural networks. Int. J. Control. **96**(1), 1–17 (2021). https://doi.org/10.1080/00207179.2021.1978555
24. Zak, M., Park, H.: Gray-box approach to fault diagnosis of dynamical systems. In: 2001 IEEE Aerospace Conference Proceedings (Cat. No. 01TH8542), vol. 2, pp. 2–669. IEEE (2001)
25. Aivaliotis, P., Arkouli, Z., Georgoulias, K., Makris, S.: Degradation curves integration in physics-based models: towards the predictive maintenance of industrial robots. Robot. Comput.-Integr. Manufact. **71**, 102177 (2021)

Artificial Intelligence at Manufacturing Equipment Level

Unsupervised Machine Learning for Blind Rivets Quality Inspection

Ander Martin Rebe[1(✉)], Mariluz Penalva[1], Fernando Veiga[1,2], Alain Gil Del Val[1,3], and Bilal El Moussaoui Abousoliman[1]

[1] TECNALIA, Basque Research and Technology Alliance (BRTA), Parque Científico y Tecnológico de Guipúzcoa, 20009 Donostia-San Sebastián, Spain
{ander.martin,cooperacion-110811,mariluz.penalva,fernando.veiga, alain.gil}@tecnalia.com, alain.gildelval@unir.net

[2] Departamento de Ingeniería, Universidad Pública de Navarra, Edificio Departamental Los Pinos, Campus Arrosadía, 31006 Pamplona, Navarra, Spain

[3] International University of La Rioja UNIR, Logroño, La Rioja, Spain

Abstract. Fastening plays a crucial role in aircraft manufacturing, and the demand for automated solutions has grown. Blind rivets are appealing for automation but require indirect assessment of the formed head for quality monitoring. Unsupervised machine learning holds potential for blind rivet inspection and extends to industrial data clustering/classification. In this context, labeling industrial data is challenging due to production focus and the need for NO OK labels. Unsupervised machine learning and advanced data analysis methods offer opportunities to optimize quality control processes without manual labeling or costly experiments.

This paper proposes two approaches to address the issue by clustering time-dependent signals in the riveting process. After preprocessing the signals, different clustering techniques are applied to time-series and signal features to obtain OK and NO OK installation clusters. The first approach, using Euclidean distance and Dynamic Time Warping, yields poor clustering results. The second approach involves feature extraction using time domain and expert descriptors, along with dimensional reduction techniques (PCA, UMAP), followed by clustering techniques. UMAP combined with DBSCAN clustering achieves interesting results, with high precision and accuracy values (above 0.8) for both OK and NO OK clusters.

Keyword: Riveting · Quality Monitoring · Time-series clustering

1 Introduction

Riveting is a widely used method for fastening components in various industries, including aerospace, automotive, and construction [1]. It offers strong and durable joints. Among the different types of rivets, blind rivets stand out as a versatile and efficient option, particularly in aerospace applications.

Mechanical fastening with rivets has long been recognized as a rapid and cost-effective solution for joining components. It involves the insertion of a rivet into pre-drilled holes and the subsequent deformation of the rivet's tail to form a secure joint.

A. Wagner et al. (Eds.): ESAIM 2023, LNME, pp. 73–80, 2024.
https://doi.org/10.1007/978-3-031-57496-2_8

Riveting has been extensively employed in airframe assembly, where it plays a vital role in joining structural components such as panels, frames, and stringers. The use of rivets ensures structural integrity, reliability, and ease of maintenance in aerospace structures [2].

Blind rivets, in particular, offer distinct advantages in terms of their installation process. Unlike solid rivets that require access from both sides of the assembly, blind rivets can be installed from the front side alone. This characteristic makes blind rivets especially suitable for applications where accessibility to the backside of the joint is limited or restricted [3]. Automating blind rivet inspection is a critical challenge in the aerospace industry. Zhu X.B. [4] has developed a classification and anomaly detection method for diferent kind of rivets based on images. The quality and reliability of blind rivet installations are paramount for ensuring structural performance and safety. Traditional inspection methods, relying on manual visual examination and verification, are time-consuming, labor-intensive, and prone to errors. To improve productivity, cost-efficiency, and overall assembly efficiency, automating the inspection process is essential.

Work is being done to automate defect detection, one example is the defect tracking tool framework for multi-process products [5].

The primary difficulty in automating blind rivet inspection lies in evaluating the deformed head on the inaccessible backside of the assembly. Unlike solid rivets that can be directly inspected, blind rivets require indirect assessment methods due to limited access to the joint's underside. Consequently, there is a need for innovative and dependable inspection techniques that can accurately evaluate the quality of blind rivet installations.

Addressing this challenge requires a combination of advanced sensing technologies, data-driven approaches, and automation concepts [6, 7]. Sensor-based monitoring systems offer a promising avenue for indirect inspection without the limitations associated with traditional non-destructive testing techniques [8, 9]. These systems can capture and analyze various process variables, such as torque and angle to infer the quality of blind rivet installations.

2 Methods and Results

Based on the paper On-Line Monitoring of Blind Fastener Installation Process by Javier Camacho et al. [10], this paper tries to apply the results to more realistic industrial cases. This work aims to separate the data into clusters of well-installed rivets and defective rivets using unsupervised machine learning techniques. This was achieved in the mentioned paper, but extending the methodology to more realistic data, introducing the source of error in terms of material thickness or chosen rivet dimension, did not achieve these good results, results are limitated by the conditioned data that was used.

The data used for this work are the same as those used by Alain Gil del Val et al. in Quality monitoring of blind fasteners installation: An approach from the manufacturing chain and visual analytics [11], with the same pre-processing.

A first step for this work has been to understand the signals by plotting them and trying to find features that describe the behaviour of the rivets. Some interesting conclusions have been drawn from this analysis. The installation diagram shows the relationship between the applied torque and the rivet rotation, providing an indirect view of the steps involved in the fastening process. Figure 1 presents an example of a blind riveting diagram. Installation begins after inserting the rivet into the hole and seating the nut in the countersink. When the drive nut starts to rotate, the sleeve and the screw rotate together, causing the sleeve to advance towards the underside as the torque is increased. When the sleeve stops rotating, the maximum torque is reached. At this point, the sleeve continues its approach to the lower face, accompanied by a decrease in torque. On reaching the bottom face, the sleeve deforms, causing a rapid increase in torque as the deformation enters the plastic regime. Finally, the bolt is pulled out during the deformation of the sleeve and breaks. The blind head deforms completely, indicating the completion of the installation process.

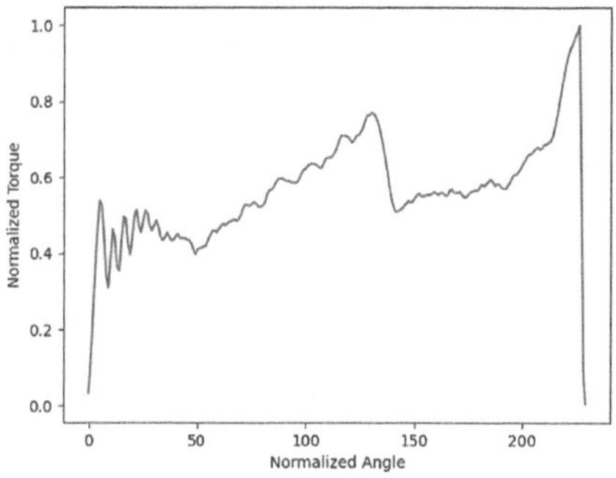

Fig. 1. Normalized Torque-Angle installation diagram.

The behavior shape of most rivets is similar, but the differences between signals do not indicate variations in quality or differences between recommended and installed rivet sizes. Notably, differences were observed among rivets from different batches, but installed under the same conditions. Therefore, the primary signal differences are not associated with installation conditions. However, the quality of the installation is indeed influenced by the installation conditions, specifically the variance between the recommended rivet size for a given thickness and the actual installed rivet size.

According to Fig. 1 and the description made of the process before, it has 5 phases, but is not easy to find the start and the end of each phase, because comparing different rivets installations there is an offset between phases as they don't always take the same portion of the process. These results have led the study to the following two methods:

2.1 Method Based on Whole Time Series

Fig. 2. Flow chart of method based on whole time series.

First approach is to work considering each and every one of the points of the riveting signal (torque-rev), after the truncation of the data, comparing the full truncated signals. For that different metrics as similarity measure are available, in this work Euclidean distance and dynamic time warping (DTW) have been tried. Figure 3 and Fig. 4 show the performance of each one. The dissimilarity is represented as the sum the orange distances for each pair of signals creating a matrix of dissimilarity (Fig. 2).

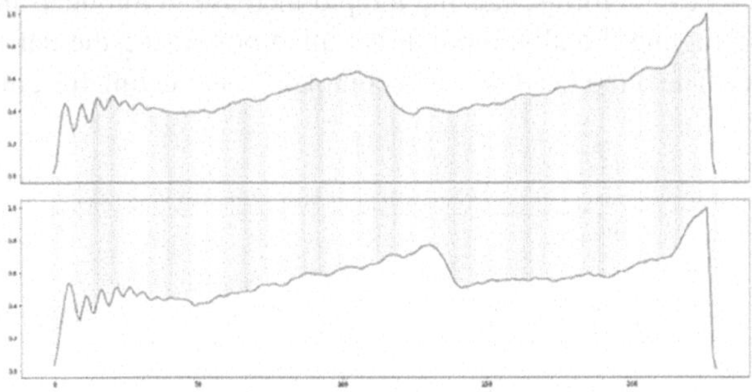

Fig. 3. Euclidean distance.

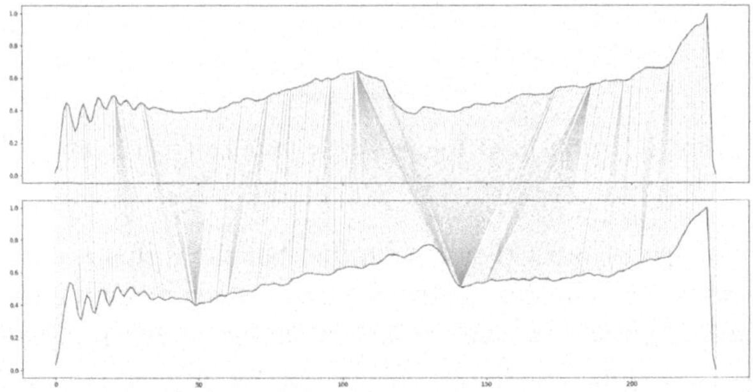

Fig. 4. Dynamic time warping distance.

Several well-known clustering algorithms were applied to the data using the similarity measures mentioned. Some algorithms from the scikit-learn library were used, which is a widely used open-source machine learning library for Python. Additionally, clustering algorithms from specialized libraries such as Tslearn, DTAIDistance, and Pyclustering were also explored. The specific clustering techniques tried are as follows (Table 1):

Results for this approach were not good. In found clusters OK and NOK samples were mixed. The techniques that need the number of clusters as input were not able to divide rivets between OK and NOK. With DBSCAN (Density-Based Spatial Clustering of Applications with Noise), a popular density-based clustering algorithm used to group together data points based on their density in a feature space, some different and a bit better results were obtained. After several trials tuning parameters, clusters with only OK and clusters with only NOK measurements have been obtained. Nevertheless, a

Table 1. Tried clustering techniques.

LIBRARY	CLUSTERING ALGORITHM
Scikit-learn	Spectral clustering
Scikit-learn	Agglomerative clustering (linkage: complete)
Scikit-learn	Agglomerative clustering (linkage: average)
Scikit-learn	Agglomerative clustering (linkage: single)
Scikit-learn	DBSCAN
Tslearn	Time series Kmeans
DTAIDistance	K-medoids
pyclustering	K-medoids

substantial portion of the acceptable measurements are intermingled within clusters of unacceptable measurements, thereby yielding unfavorable outcomes.

2.2 Method Based on Feature Extraction

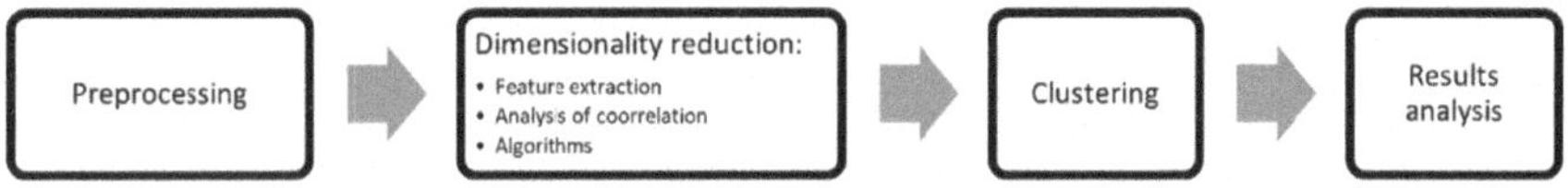

Fig. 5. Flow chart of method based on feature extraction.

Dimensionality reduction plays a crucial role in the field of machine learning and data analysis by addressing the challenge of high-dimensional data that causes an increased computational costs, overfitting, and difficulty in interpreting the data (Fig. 5).

In this work 3 steps are applied for reducing the amount of data: first, feature extraction; second, analysis of correlation focused on the aim of the work and; third, by dimensionality reduction algorithms that offer a powerful solution by reducing the number of features while preserving important information.

Feature extraction:

This approach is effective for clustering time series due to two main reasons. Firstly, it utilizes descriptive indicators, known as descriptors, to summarize the key characteristics of a signal, reducing data and computation requirements. Secondly, these descriptors are linked to the physical aspects of the process, aiding in understanding the process's physics and identifying the root causes of failures. The descriptors used in this work can be categorized as domain expert descriptors [11].

Analysis of correlation focused on the aim of the work:

All extracted features are not related with the failure mode of blind rivets, so before applying dimensionality reduction algorithms, an analysis of correlation between features and quality descriptor has been developed. Due to that, the features more related with the quality of the riveting are the following:

- The maximum value of the signal at phase 4.
- The minimum value and Area under the Curve of the signal at phase 4 and 5.
- The coefficients 'a' and 'b' in the equation of a line (ax + b) that best fits the signal during phase 5 and 6.
- Error Correction Model (ECM) of the signal during phase 5 to the equation of a line that best fits.
- The area under the curve of the signal during 3rd, 4th and 5th equal segments with a 10% of overlapping.

It is interesting and makes complete sense, according to the process, that the final states of the process are the most related with the failure of the installation. After the correlation analysis, risk of losing relevant information related with failure by applying dimensionality reduction algorithms is less.
Dimensionality reduction algorithms:

In this work PCA (Principal Component Analysis) or UMAP (Uniform Manifold Approximation and Projection) have been tried. After the dimensionality reduction the following results are obtained (Fig. 6):

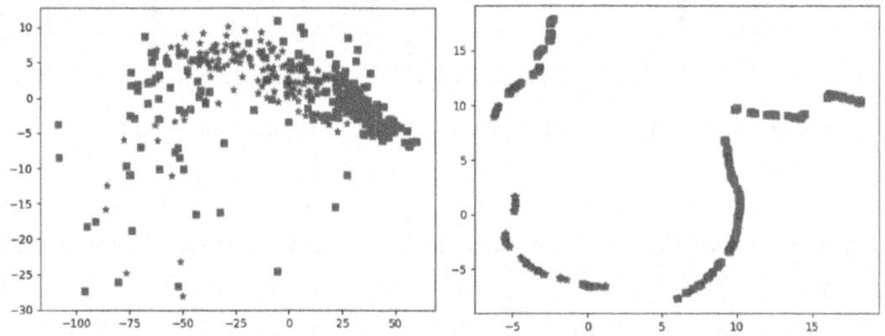

Fig. 6. PCA (left) and UMAP (right) 2-D distribution

The pictures show the new 2-dimensional distribution of the samples on principal component axis in the picture on the left and after UMAP dimensionality reduction on the right. Red points are failed rivets and green ones represent ok installed rivets.

3 Clustering Results and Discussion

The results of UMAP dimensionality reduction technique, see Fig. 7, show that there are visual groups of samples in which is possible to divide using DBSCAN method, based on the density of samples. The application of UMAP to pertinent features significantly enhances the outcomes of DBSCAN clustering (Table 2).

The accuracy and precision metrics point out a high percentage of correct labels for both OK and NOK installations. On the other hand, the Silhouette score and ARI values

Table 2. Clustering Results

Cluster	Samples	OK	NO OK
1	34	0	**34**
2	88	5	**83**
3	17	0	**17**
4	31	0	**31**
5	45	**24**	21
6	15	2	**13**
7	25	7	**18**
8	76	**67**	9
9	17	**14**	3
10	22	**21**	1
11	13	3	**10**
12	17	6	**11**

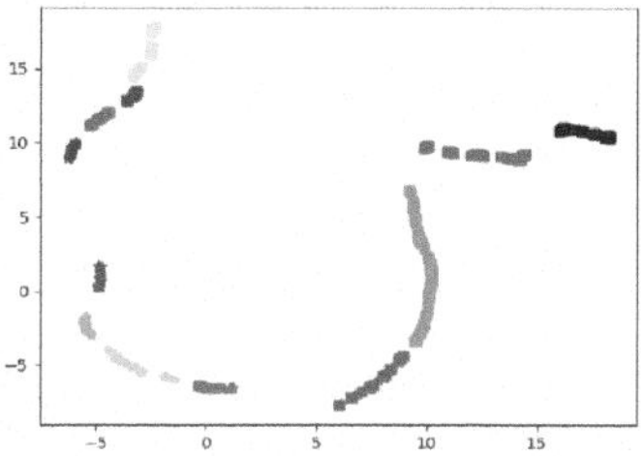

Fig. 7. Visual clustering Results.

confirm the complexity of the classification. Though installations must be classified just into OK/NOK, multiple variables affecting the installation outcome lead to a high number of groups, making clustering more challenging.

In these results 12 clusters are obtained, 4 with a majority of OK rivets and 8 with a majority of NO OK rivets. There are 3 clusters with NO OK installations exclusively.

The results of the clustering model are insufficient for accurately detecting OK installations due to the achieved precision. However, it remains valuable for estimating the quality of an installation by effectively identifying NO OK installations, where errors have a lesser impact.

Every time groups with almost 100% accuracy were obtained, the characteristics that unite them in the same group were not perceived. This gives value to the result as hidden patterns have been found thanks to clustering. "Large selected" and "on air" are the most clearly grouped NOK installations in the first 4 clusters, this is relevant because this type of defects is of major concern since in practise they represent a loose joint.

4 Conclusion

Unsupervised machine learning techniques and advanced data analysis methodologies can optimize quality control processes without the need for manual labeling or costly trial and error experiments.

The first approach, did not yield satisfactory clustering results. The second approach, showed more promising results and can be effective for blind rivet quality inspection.

The proposed methods have potential applications in industrial settings where labeling or classification of non-labeled data is challenging, and companies can optimize their quality control processes without extensive manual efforts or trial and error experiments. Unsupervised methods identify patterns and structures in the data without the need for labeled examples.

Funding. The authors acknowledge the funding from the European Union's Horizon 2020 research and innovation programme under grant agreement No. 958303.

References

1. Maggiore, S., Banea, M.D., Stagnaro, P., Luciano, G.: A review of structural adhesive joints in hybrid joining processes. Polymers **13**, 3961 (2021). https://doi.org/10.3390/polym13223961
2. Tavares, S.M.O., de Castro, P.M.S.T.: An overview of fatigue in aircraft structures. Fatigue Fract. Eng. Mater. Struct. **40**, 1510–1529 (2017). https://doi.org/10.1111/ffe.12631
3. Amancio-Filho, S.T., dos Santos, J.F.: Joining of polymers and polymer–metal hybrid structures: recent developments and trends. Polym. Eng. Sci. **49**, 1461–1476 (2009). https://doi.org/10.1002/pen.21424
4. Zhu, X.: Aviation rivet classification and anomaly detection based on deep learning. Int. J. Aerosp. Eng. **2023**, 3546838 (2023). https://doi.org/10.1155/2023/3546838
5. Papacharalampopoulos, A., Petrides, D., Stavropoulos, P.: A defect tracking tool framework for multi-process products. Procedia CIRP **79**, 523–527 (2019). https://doi.org/10.1016/j.procir.2019.02.100
6. Pech, M., Vrchota, J., Bednář, J.: Predictive maintenance and intelligent sensors in smart factory: review. Sensors **21**, 1470 (2021). https://doi.org/10.3390/s21041470
7. Lomas, M.C., Beldarrain, T.O., Canas, G.R., Suarez, F.V., Oscoz, M.P., Mentxaka, A.L.: Mejora del mecanizado mediante el registro y análisis de variables de proceso en la nube. DYNA **95**, 472–477 (2020). https://doi.org/10.6036/9398
8. Wang, W.-M., Ali Khan, H., Li, J., Miller, S.F., Zachary Trimble, A.: Classification of failure modes in friction stir blind riveted lap-shear joints with dissimilar materials. J. Manufact. Sci. Eng. **139**, 021005 (2016). https://doi.org/10.1115/1.4034280
9. Gao, Z., Khan, H.A., Li, J., (Grace) Guo, W.: Sensor fusion and on-line monitoring of friction stir blind riveting for lightweight materials manufacturing. J. Manufact. Sci. Eng. **144**, 061009 (2021). https://doi.org/10.1115/1.4052907
10. Camacho, J., Veiga, F., Penalva, M.L., Diez-Olivan, A., Deitert, L., López de Lacalle, N.: On-line monitoring of blind fastener installation process. Materials **12**, 1157 (2019). https://doi.org/10.3390/ma12071157
11. Del Val, A.G., Penalva, M., Veiga, F., Iriondo, E.: Quality monitoring of blind fasteners installation: an approach from the manufacturing chain and visual analytics. IFAC-PapersOnLine. **55**, 270–276 (2022). https://doi.org/10.1016/j.ifacol.2022.04.205

Enabling Seamless Human-Robot Collaboration in Manufacturing Using LLMs

Christos Gkournelos, Christos Konstantinou, Panagiotis Angelakis, George Michalos, and Sotiris Makris(✉)

Laboratory for Manufacturing Systems and Automation (LMS), Department of Mechanical Engineering and Aeronautics, University of Patras, 26504 Patras, Greece
makris@lms.mech.upatras.gr

Abstract. In the era of Industry 5.0, there is a growing interest in harnessing the potential of human-robot collaboration (HRC) in manufacturing environments. This research paper focuses on the integration of Large Language Models (LLMs) to augment HRC capabilities, particularly in addressing configuration issues when programming robots to collaborate with human operators. By harnessing the capabilities of LLMs in combination with a user-friendly augmented reality (AR) interface, the proposed approach empowers human operators to seamlessly collaborate with robots, facilitating smooth and efficient assembly processes. This research introduces the CollabAI an AI assistant for task management and natural communication based on a fine-tuned GPT model focusing on collaborative manufacturing. Real-world experiments conducted in two manufacturing settings coming from the automotive and machinery industries. The findings have implications for various industries seeking to increase productivity and foster a new era of efficient and effective collaboration in manufacturing environments.

Keywords: Artificial Intelligence · Human-Robot Collaboration · Large Language Models

1 Introduction

In the era of Industry 5.0 (I5.0) [1], where human-robot collaboration (HRC) [2] is becoming increasingly prevalent in manufacturing environments, the demand for mutual and intuitive interaction between humans and robots has never been more critical. In comparison with I4.0 where AI was initially introduced in Manufacturing [3, 4], I5.0 emphasizes human-robot reconciliation under the terms of collaboration of AI agents with human agents, while preserving social values. Trust is considered a crucial component in fostering collaborative relationships, as it has a significant influence on both the level of collaboration and productivity [5]. Prior studies have investigated the utilization of natural language for controlling robots [6, 7]. However, there are still challenges to overcome in terms of their ability to understand task context, engage in logical reasoning and facilitate a "human-like" communication.

A. Wagner et al. (Eds.): ESAIM 2023, LNME, pp. 81–89, 2024.
https://doi.org/10.1007/978-3-031-57496-2_9

To achieve seamless collaboration with an autonomous agent Sugiura et al. [8] presented a grounded instruction understanding for social robots based on a custom Generative Adversarial Nets (GAN) classifier, Yang Y. et al. [5] and Kouba A. [9] have both presented a solution that incorporates the ChatGPT cloud and offers an interface in Robotic Operating System (ROS) [10]. Kouba's work focuses on the conversion of unstructured data from human speech into structured commands that can be parsed from a robot as direct commands. Yang Y. et al. investigate the effect on trust of using an intelligent Large Language Model (LLM) for robotic control in an HRC assembly task. Despite the success of LLMs in certain applications, current solutions are limited in their understanding of manufacturing systems' inner workings. The key causes of these limitations are that a) the available models lack access to industry-specific proprietary information [11] and b) they lack access to the current state of the system.

HRC production systems can exhibit complexity in their data structures, requiring sophisticated frameworks to manage and process vast amounts of information [12]. Additionally, ensuring user-friendliness for non-expert users presents a challenge, as the systems need to be designed with intuitive interfaces and simplified interactions to enable easy involvement and participation [13]. Balancing the intricacies of data handling and user accessibility is vital for developing successful collaborative production systems that effectively integrate both technical complexity and user-friendly interfaces.

While LLMs serve as the essential backbone of any future AI system, they are not without their shortcomings. LLMs' tendencies to perpetuate biases present a significant concern, and the generation of misleading or false information is a notable drawback. The issue then becomes how to fully exploit the advantages of self-supervised deep language models (LMs) while mitigating their disadvantages and ensuring the highest ethical standards. This study proposes CollabAI, a scalable and explicable AI assistant for collaborative manufacturing applications. CollabAI seeks to resolve the complexity in task management of multiple collaborative resources (robots, operators) and increase the mutual understanding between human and artificial agents with a simple communication interface.

2 Approach

This work presents an LLM application called CollabAI, facilitating mutual and intuitive interaction between humans and robots in the context of I5.0. The proposed solution features a completely flexible architecture, and the high-level design is depicted in Fig. 1. CollabAI consists of an extendable set of modules. These modules can be:

- **Agents**: Neural modules including LLMs, fine-tuned LMs, or any other AI-enabled component with the objective to classify an action based on inputs.
- **Tools**: Symbolic modules that implement action functions called by Agents. These modules are used for connecting external APIs, such as databases, robot controllers, etc.

The core input/output functionalities of CollabAI modules can be categorized into *Format*, *Predict,* and *Parse*. Format modules can be either Tools or Agents and their responsibility is to templatize and manage the model inputs based on prompts. Agents

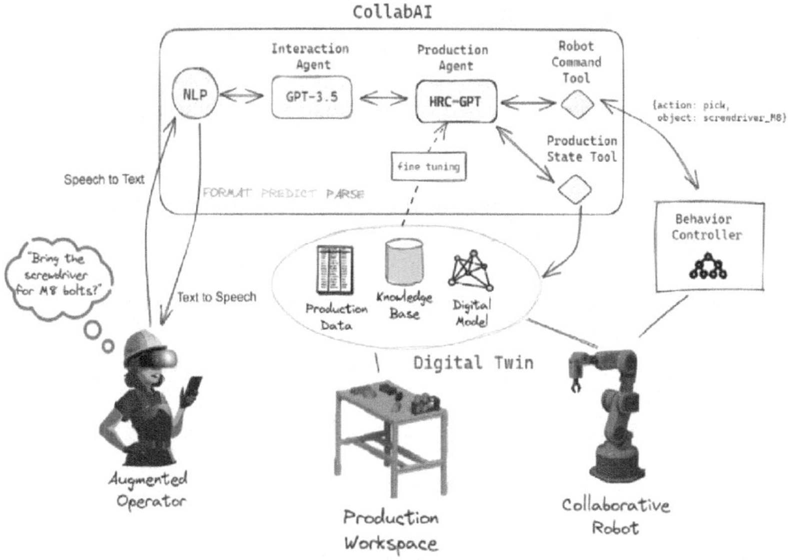

Fig. 1. CollabAI general architecture

have predictive responsibilities based on their language model. Parse functionalities are provided by Tool modules that extract information from the Agents' model outputs to execute an action.

CollabAI aims to provide the following benefits compared to existing LM applications:

1. System knowledge: Access to manufacturing proprietary knowledge bases and data sources for fine-tuning Agent's LLMs according to each manufacturing case.
2. Online information: The integration with external production status monitoring components, allows the CollabAI to retrieve dynamically updated information for correctly responding to inputs that static models cannot.
3. Flexible scalability: This modular architecture allows easy reconfiguration and addition of the Agent and Tool components based on application requirements. Moreover, the training of the fine-tuned Agents is performed independently to easily add new capabilities without compromising the performance of the existing ones.
4. Explainability: Agents are programmed based on the chain-of-thought [14] paradigm, where they receive input, plan a sequence of thought steps, and finally execute these steps. This transparent process can be monitored by the users, knowing all the observations at every step. This provides a level of understanding and control over the AI system's behavior.

Human-Robot collaborative production systems require mutual communication for efficient synergy. Following the paradigm of seamless HRC, a Digital Twin component [12] is responsible for handling the data modeling of the production workspace and controlling the behavior of collaborative robots (cobot). CollabAI has the required interfaces for connecting with the production Digital Twin and retrieving information for LM fine-tuning and cobot behavior controlling.

The interaction with the human operators is handled by a Natural Language Processing (NLP) Agent. This Tool module is responsible for formatting the user's speech or text input. Also, as feedback to the user the NLP Agent provides the CollabAI observations both in text and in speech format. This Agent can be easily integrated with AR/VR applications, increasing the immersion of the user.

This work will analyze the implementation of the CollabAI consisting of one formatting Agent with NLP capabilities, two LM Agents (Interaction Agent and Production Agent), and two action Tools (Robot Command, Production Status). This topology is designed based on the HRC requirements that were identified in the case study.

3 Implementation

3.1 Agents

CollabAI utilizes neural modules (agents) and based on the operator's input makes observations of the system. Agents in CollabAI employ a diverse set of tools and dynamically select their applications.

The NLP Agent is responsible for providing an oral communication interface between human operators and autonomous resources. This module is deployed based on Azure's speech-to-text [15] framework that is running on HoloLens augmented reality glasses.

The Interaction Agent facilitates the initial level of communication, engaging in general-purpose discussions with operators. It operates on a plan-and-execute strategy and utilizes a conversational memory buffer. To meet the real-time interaction requirement, a series of tests were conducted to select the most suitable LLM for the Interaction Agent. Several models, including *Groovy v1.3*, *MPT-7B*, and *Snoozy-13B*, known for their reasoning abilities and accurate output, were evaluated. The models were tested with an equal number of input tokens (ranging from 5 to 250), and the maximum response times were set at 10, 50, and 300 s respectively. Based on the results, *GPT-3.5* over OpenAI cloud was chosen for its impressive response time of 3 s. This selected model has been fine-tuned using prompting techniques to distinguish whether the input requires redirection to the Production Agent or if it pertains to a general-purpose question.

The Production Agent addresses questions related to the HRC assembly deployed in the system. To effectively handle these inquiries, a custom Large Language Model (LLM) named HRC-GPT was developed based on the text-DaVinci-003 model. The training process, illustrated in Fig. 2, comprises two distinct phases. Firstly, the indexing method is employed to transform data from external sources such as files and databases into a structured format that can be easily queried by the language model. Subsequently, in the training phase, the external data in a structured format is utilized along with a standard *GPT* model to generate a substantial set of prompts. Using these generated prompts, the HRC-GPT model performance and adaptability are enhanced. As stated in [16] LLMs such as GPT can be considered "few-shot learners". It is proven, that huge LMs, perform better in task-agnostic tasks using paradigms as input training examples.

In our scenario, they can efficiently adapt their knowledge base, using such prompt examples. These examples were specifically designed to encapsulate the HRC context containing the production data information and the system's knowledge base.

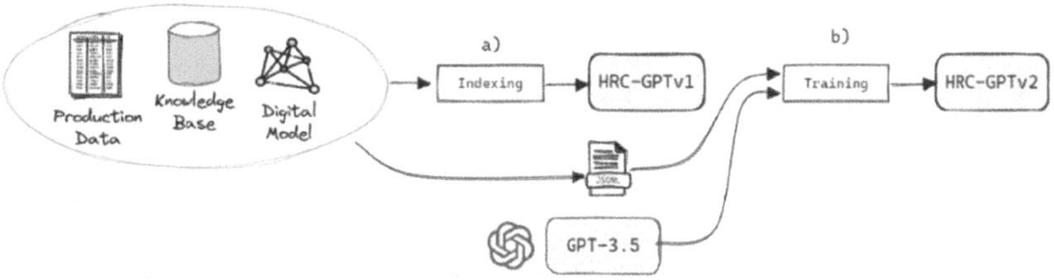

Fig. 2. Training procedures for the Production Agent, a) Data indexing, and b) OpenAI custom model fine-tuning

3.2 Tools

The CollabAI solution incorporates two essential tool modules for connecting with the physical world: the Production State Tool (PST) and the Robot Command Tool (RCT). Within the current approach, the Production Agent serves as a tool the Interaction Agent and interfaces with two other tools. If a simple user command of the type "pause," "resume," "abort," or "go home" is received from the Interaction Agent, it invokes the Robot Command Tool (RCT). In the case of a question regarding the knowledge base or process, the Production Agent calls the Production State Tool (PST) and subsequently determines whether to involve the RTC before returning the observation or status of the command to the Interaction Agent.

The RCT tool forwards its input to the reactive Robot Behavior Controller, which possesses internal knowledge derived from the Production Data on how to perform each task. The Robot Behavior Controller can safely preempt tasks before executing new ones and prompt the user for recovery in case of errors.

The Production State Tool (PST) retrieves current information from the production system based on the input provided by the Production Agent. It can also modify the system's state if the user identifies any discrepancies between the Digital Twin and the real cell.

Both the RTC and PST tools utilize JSON format for constructing their input and output. They return observations to the Production Agent and employ ROS service requests to communicate with external components and physical hardware.

3.3 System Deployment

The CollabAI solution comprises several essential external components that work in tandem. The sequence of interactions between the HRC components is illustrated in Fig. 3. Starting from the human perspective, the primary user interface of the system is Hololens2, an AR headset. This device serves as both an input and output interface, capable of converting speech to text through the NLP module and rendering visual instructions. Communication between the system components, including the Digital Twin and Robot behavior, is facilitated via the ROS API, with the tool modules taking charge of the interfacing process.

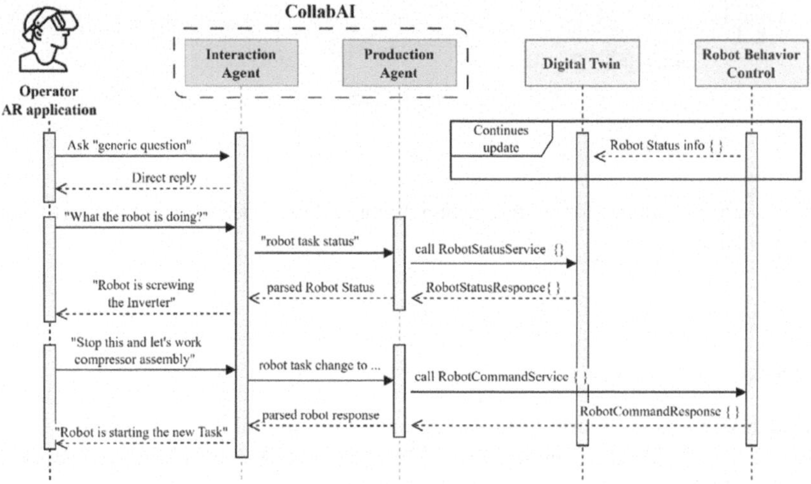

Fig. 3. Sequence diagram of Human-Robot collaboration with CollabAI

4 Case Study

CollabAI has been implemented and thoroughly evaluated in two distinct manufacturing use cases. The first use case originates from the automotive industry and involves the assembling of an electric motor (e-motor) and its inverter, while the second use case comes from the machinery industry and involves the assembly of an industrial air compressor. The selection of two cases was intended to demonstrate the adaptability of CollabAI to be readily adapted to various manufacturing sectors.

As seen in Fig. 4, the reasoning behind the thoughts of the agents working together gives a result that follows a basic logic conclusion. The CollabAI system is designed to simplify user queries, allowing the main interaction to be in natural human language. By streamlining the user's inquiries, CollabAI can reach conclusive outcomes efficiently and effectively that can communicate with the autonomous resources in the collaborative workstation.

Figure 5 depicts the collaborative workstation utilized for testing CollabAI. The workstation's design aims to create an environment conducive to assembly tasks, promoting ample opportunities for potential collaboration between humans and robots. By leveraging the operator's input and the knowledge stored in the Digital Twin, CollabAI utilizes reasoning capabilities to effectively control the behavior of the cobot during task execution.

```
[Operator]: 'Ok, lets start the assembly of the electric motor'
[Interaction Agent]: [Thought]: 'The user wants to assemble the electric motor'
[Interaction Agent]: [Action]: 'ProductionSearch'
[Interaction Agent]: [Action Input]: 'Inverter'
   [Production Agent]: [Thought]: 'What are the parts needed to assemble the inverter?'
   [Production Agent]: [Action]: 'RobotStateSearch'
   [Production Agent]: [Action Input]: 'Parts needed to assemble the inverter'
[Interaction Agent]: [Final Answer]: 'Engine1/Inverter'
[Interaction Agent]: [Thought]: 'The user wants to start screwing the screws'
[Interaction Agent]: [Action]: 'ProductionSearch'
[Interaction Agent]: [Action Input]: 'Screws'
   [Production Agent]: [Thought]: 'What type of screws are needed for Engine1/Inverter?'
   [Production Agent]: [Action]: 'RobotStateSearch'
   [Production Agent]: [Action Input]: 'The screws needed for
   Engine1/Inverter are Engine1/Screw01, Engine1/Screw02, Engine1/Screw03, Engine1/Screw04 ..'
[Interaction Agent]: [Final Answer]: To assemble the electric motor, you will
need the Engine1/Inverter and the screws Engine1/Screw01, Engine1/Screw02, Engine1/Screw03
```

Fig. 4. CollabAI response and chain of thoughts

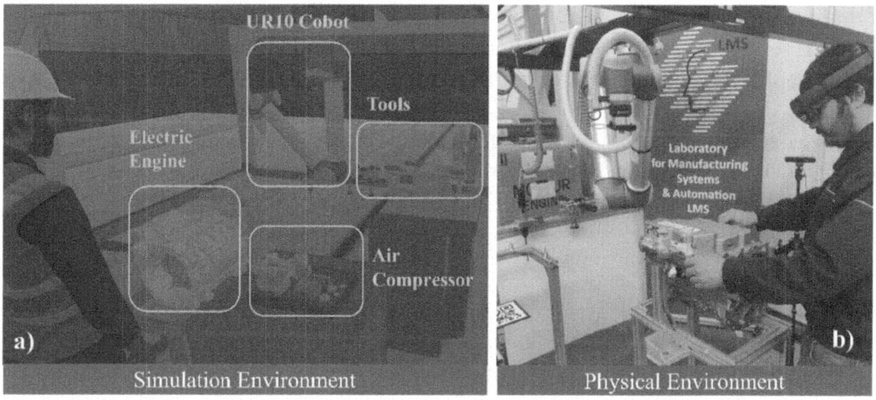

Fig. 5. Human-Robot collaborative assembly a) Simulation b) Physical Environment

5 Conclusions and Future Work

In conclusion, the CollabAI framework stands as a significant advancement in the realm of Human-Robot Collaboration (HRC) within the context of Industry 5.0. By facilitating seamless interaction between non-expert operators and complex robotic systems through natural language communication, CollabAI holds the potential to revolutionize collaborative processes in manufacturing environments. The modular architecture of CollabAI not only ensures adaptability but also enables swift integration into diverse operational scenarios, catering to the needs of technical stakeholders.

While the preliminary outcomes of CollabAI are encouraging, there remains a critical avenue for further exploration through user studies and soliciting feedback from both operators and non-expert users. These endeavors are anticipated to provide invaluable insights, which will be instrumental in shaping the direction of our future research. The implications of CollabAI's findings extend beyond mere technological innovation; they have the power to reshape productivity paradigms across various industries, ushering in a new era of streamlined and effective collaboration within manufacturing ecosystems.

Examining the approach's potential limitations, it is important to address certain factors that warrant consideration. For instance, the round-trip delay associated with human commands merits discussion. The impressive 3-s processing time, as highlighted in the sequence diagram, is indeed noteworthy, yet it's essential to acknowledge that real-world application necessitates multiple prompts—resulting in potential cumulative delays. Moreover, the use of CollabAI in real production environments raises pertinent questions about its adaptability to varying operational tempos. While the system's capabilities are evident, navigating scenarios where response time is a critical factor poses a distinctive challenge. Balancing rapid decision-making with complex tasks demands a nuanced strategy that could encompass pre-emptive processing, predictive analytics, and adaptive machine learning models.

In summary, CollabAI's implications for Industry 5.0 are profound, promising heightened collaboration between humans and robots. Our envisioned trajectory involves delving deeper into user-centric evaluations, refining response time dynamics, and fortifying the system's learning capacities. This journey, guided by the understanding that CollabAI serves as a bridge between human and machine intelligence, holds the potential to reshape the very foundations of collaborative manufacturing processes.

Acknowledgments. This research has been supported by the EC funded project "ASSISTANT: Learning and robust decision Support systems for agile manufacturing environments" (Grant Agreement: 101000165).

References

1. Yavari, F., Pilevari, N.: Industry revolutions development from Industry 1.0 to Industry 5.0 in manufacturing. J. Ind. Strateg. Manag. **5**(2), 44–63 (2020)
2. Makris, S.: Cooperating Robots for Flexible Manufacturing. In: Springer Series in Advanced Manufacturing. Springer International Publishing, Berlin, Heidelberg (2021). https://doi.org/10.1007/978-3-030-51591-1
3. Chryssolouris, G., Alexopoulos, K., Arkouli, Z.: A Perspective on Artificial Intelligence in Manufacturing. In: Studies in Systems, Decision and Control. Springer International Publishing, Cham (2023). https://doi.org/10.1007/978-3-031-21828-6_4
4. Makris, S., et al.: ARTIFICIAL INTELLIGENCE IN MANUFACTURING White paper Prepared by the Artificial Intelligence in Manufacturing Network -AIM-NET. May (2023)
5. Ye, Y., You, H., Du, J.: Improved Trust in Human-Robot Collaboration with ChatGPT (2023). arXiv. https://doi.org/10.48550/arXiv.2304.12529
6. Gkournelos, C., Karagiannis, P., Michalos, G., Makris, S., Kousi, N., Koukas, S.: Application of wearable devices for supporting operators in human-robot cooperative assembly tasks. In: Procedia CIRP, pp. 177–182 (2018). https://doi.org/10.1016/j.procir.2018.01.019
7. Michalos, G., et al.: Seamless human robot collaborative assembly – an automotive case study. Mechatronics **55**, 194–211 (2018). https://doi.org/10.1016/J.MECHATRONICS.2018.08.006
8. Sugiura, K., Kawai, H.: Grounded language understanding for manipulation instructions using GAN-based classification. In: 2017 IEEE Automatic Speech Recognition and Understanding Workshop (ASRU), Okinawa, Japan, pp. 519–524. IEEE (2017). https://doi.org/10.1109/ASRU.2017.8268980

9. Koubaa, A.: ROSGPT: Next-Generation Human-Robot Interaction with ChatGPT and ROS (2023). Preprints. https://doi.org/10.20944/preprints202304.0827.v2
10. Quigley, M., et al.: ROS: an open-source Robot Operating System. In: International Conference on Robotics and Automation (2009). http://stair.stanford.edu
11. Loureiro, D., Barbieri, F., Neves, L., Anke, L.E., Camacho-Collados, J.: TimeLMs: Diachronic Language Models from Twitter (2022). arXiv. https://doi.org/10.48550/arXiv.2202.03829
12. Kousi, N., Gkournelos, C., Aivaliotis, S., Giannoulis, C., Michalos, G., Makris, S.: Digital twin for adaptation of robots' behavior in flexible robotic assembly lines. Procedia Manufact. **28**, 121–126 (2019). https://doi.org/10.1016/j.promfg.2018.12.020
13. Wang, L., et al.: Symbiotic human-robot collaborative assembly. CIRP Ann. **68**(2), 701–726 (2019). https://doi.org/10.1016/j.cirp.2019.05.002
14. Wei, J., et al.: Chain-of-Thought Prompting Elicits Reasoning in Large Language Models (2023). arXiv. https://doi.org/10.48550/arXiv.2201.11903
15. Microsoft Azure, "Speech to Text – Audio to Text Translation. https://azure.microsoft.com/en-us/products/cognitive-services/speech-to-text. Accessed 06 Jul 2023
16. Brown, T.B., et al.: Language Models are Few-Shot Learners (2020). arXiv. http://arxiv.org/abs/2005.14165. Accessed 09 Jun 2023

Enhancing Object Detection Performance for Small Objects Through Synthetic Data Generation and Proportional Class-Balancing Technique: A Comparative Study in Industrial Scenarios

Jibinraj Antony[1(✉)], Vinit Hegiste[2], Ali Nazeri[1], Hooman Tavakoli[1], Snehal Walunj[1], Christiane Plociennik[1], and Martin Ruskowski[1,2]

[1] German Research Center for Artificial Intelligence (DFKI), Kaiserslautern, Germany
jibinraj.antony@dfki.de

[2] Rheinland-Pfälzische Technische Universität (RPTU) Kaiserslautern-Landau, Kaiserslautern, Germany

Abstract. Object Detection (OD) has proven to be a significant computer vision method in extracting localized class information and has multiple applications in the industry. Although many of the state-of-the-art (SOTA) OD models perform well on medium and large sized objects, they seem to under perform on small objects. In most of the industrial use cases, it is difficult to collect and annotate data for small objects, as it is time-consuming and prone to human errors. Additionally, those datasets are likely to be unbalanced and often result in an inefficient model convergence. To tackle this challenge, this study presents a novel approach that injects additional data points to improve the performance of the OD models. Using synthetic data generation, the difficulties in data collection and annotations for small object data points can be minimized and to create a dataset with balanced distribution. This paper discusses the effects of a simple proportional class-balancing technique, to enable better anchor matching of the OD models. A comparison was carried out on the performances of the SOTA OD models: YOLOv5, YOLOv7 and SSD, for combinations of real and synthetic datasets within an industrial use case.

Keywords: Small Object Detection · Class Balancing · Synthetic Data Generation · YOLOv7 · YOLOv5 · SSD

1 Introduction

The birth of Convolutional Neural Networks (CNN) set down an important milestone in the performance improvement of image classification challenges,

The first Five Authors contributed equally to the work.

A. Wagner et al. (Eds.): ESAIM 2023, LNME, pp. 90–105, 2024.
https://doi.org/10.1007/978-3-031-57496-2_10

where networks like AlexNet [1] and their successors exceed human level performances. As the researches in CNN advanced, so did the academic interest in the localization of the specific objects, where networks like OverFeat [2] made their first success with a deep CNN architecture. Further improvements in the field of Object Detection (OD) were possible with networks like Regions with CNN features (R-CNN) [3], Fast R-CNN [4], Faster R-CNN [5], YOLO [6] and its variants. These networks and their current adaptations were the catalyst in solving some of the most challenging OD problems, enabling it to be integrated into the industry with a high degree of confidence.

	AP	AP^{50}	AP^{75}	AP^{S}	AP^{M}	AP^{L}	AR^{1}	AR^{10}
Noah CV Lab (Huawei)	0.588	0.766	0.649	0.407	0.616	0.720	0.418	0.700
mmdet	0.578	0.770	0.637	0.399	0.605	0.706	0.414	0.690
DeepAR(ETRIxKAIST_AIM)	0.553	0.746	0.609	0.378	0.583	0.668	0.403	0.674
DetectoRS	0.550	0.736	0.604	0.377	0.578	0.669	0.401	0.678
KiwiDet2	0.547	0.728	0.597	0.362	0.576	0.685	0.401	0.680
360 AI Research	0.546	0.737	0.600	0.369	0.583	0.674	0.398	0.665
ZFTurbo	0.544	0.728	0.600	0.367	0.579	0.671	0.400	0.657
Hyundai Mobis AD Lab	0.538	0.721	0.591	0.343	0.571	0.675	0.398	0.646
ByteDance_VC	0.534	0.723	0.589	0.351	0.572	0.663	0.396	0.653

Fig. 1. Detection Leader board of COCO Dataset OD Challenge from 2020. The Average Precision of Small (AP^S) and Large (AP^L) objects in the dataset for the top performing models are highlighted in the figure [7].

Detecting small objects is particularly challenging as they have very small features-space to offer for training. As per the COCO evaluation matrix [8], small objects are the objects in an image with an area less than 32×32 pixels. Although the state-of-the-art (SOTA) Machine Learning (ML) models are excellent in general OD applications involving prominently visible or dominant objects, they perform poorly in scenarios involving small objects. For instance, the latest COCO Dataset OD challenge shows that even the top performing models tend to perform badly on small objects [7] (see Fig. 1). More specifically, taking a YOLOv4 with CSPDarknet-53 backbone on MS COCO Dataset as example [9], the mean average precision (mAP) for the smaller object is only 20%, where mAP for the dominant objects are more than twice as much (mAP of 45% for the medium and 56% for the large objects), highlights the big gap in the performance of models in smaller OD problems.

However, when it comes to the integration of ML application into the industry, the data-driven nature of these approaches make them highly dependent on the data used, both quality and quantity, which is why data availability is one of the most crucial factors in a project's success. But in industrial environments, the collection of enough useful data is unfortunately a painful process, accounting for additional costs, effort, and time. Data has to be collected error-free and has to be categorized correctly in order to ensure its effectiveness in the ML application. Since the industrial machines of today are less prone to errors, the availability of the useful data necessary for the ML implementation becomes even more limited, turning the industrial realization a challenging and complex task. In some cases, the data obtained from the industrial environment is unbalanced, which consequently affects the model's performance on all *classes*. The low performance for the detection of small objects is often attributed to their unbalanced distribution over the entire dataset. A balanced dataset could eventually improve the learning of the model and hence perform better in all associated tasks.

As the manual labelling of small objects on image data is a time and effort intensive task, the latest advancements of synthetic data generation methods tend to address these difficulties in data preparation [10]. For many industrial objects, CAD data is available and can be utilized as a blue-print for the generation of synthetic data. Although CAD data lacks photo-realism, a simple rendering can be imparted and the corresponding images of the target class could be generated from their CAD models without high computational costs. Upon adding the newly generated images, the class distribution of the dataset could be balanced. This approach of leveraging synthetic data generation together with real datasets for class focused data balancing and its effectiveness in improving the performance of use cases with small OD has been addressed in this paper.

2 Background and Related Works

Object Detection is a crucial problem in computer vision that involves recognizing and localizing objects of interest within an image or a video stream [11]. In recent years, deep learning (DL) techniques using CNNs, have achieved remarkable success on improving the accuracy of OD models, even in challenging scenarios where the object features are partially occluded, poorly illuminated, or exhibit low contrast. Single Shot Detector (SSD) [12] and You Only Look Once (YOLO) [6] are two CNN-based models that have gained popularity in the field of OD. These models have various applications in transportation, military, and industrial use cases. For example, an OD model based on YOLOv2 was utilized in [13] for the surface inspection on conveyor belts, while [14] compares the effect of SSD, F-RCNN, YOLOv3 and YOLOv5 OD models in detecting surface defects in metals using a generic dataset.

Beyond general OD applications, the small OD finds its application in areal object inspection, industrial scenarios, etc. Small OD also forms the basis of many other computer vision applications such as object tracking [15], instance segmentation [16], action recognition [17] and others. The field of application of small OD is therefore diverse, where it is used for spider detection and removal application [18], in autonomous driving application [19], remote sensing [20], or even for improving the synchronization of Digital Twin of a manufacturing system [21].

Due to the data-driven nature of small OD models, they are highly dependent on the available data. In most industrial situations, collecting new datasets to train a model is a challenging, expensive and a time-consuming task [22]. These challenges could be addressed through *data augmentation*, which is also a powerful technique to alleviate over-fitting. Many researchers have proposed data augmentation methods [23,24] to artificially increase the size of training data using the data in hand without collecting new data [25]. One or more morphs are applied to the data while preserving the labels during transformation [26].

Synthetic datasets [27] can be generated based on custom requirements and controlled conditions, and shall follow the underlying distribution of the real dataset [28]. Synthetic data generation can be achieved generally via two techniques, first by using *3D rendering and simulation* software or real-time rendering frameworks [10,29] and second by using DL techniques such as Variational Autoencoder [30], Denoising Diffusion Probabilistic Models [31], GANs [32] (such as StyleGANs3) [33], etc. The most simple form of synthetic data generation is using the object masks [34], where one can have various types of background and the segmented objects of different types needed to be detected. Then a combination of these objects with the background is undertaken to create an image, i.e., to cut the object and paste it on to the background image. This method is good to train OD for some simple tasks, but would not yield better results for complex OD tasks, since this approach does not help in model generalization. While DL techniques are faster than normal cut-paste in creating a large dataset, a similar issue would arise as these models do not guarantee the desired outcome with specific angles, backgrounds, and lighting conditions. These issues can be easily tackled with synthetic data created using 3D rendering and simulation tools [35].

Usage of synthetic data for OD makes many applications easier to implement. The authors, in [36], utilize synthetic data generated using DeepGTAV framework to work on Unmanned Aerial Vehicles scenarios. While synthetic data was leveraged in [37] to create OD models for robotic object grasping application. CAD models offer an important asset for synthesizing image data. Taking advantage of the geometric precision and comprehensive annotations inherent in CAD models, it is possible to create diverse and labeled synthetic datasets that enhance the performance and generalization of OD models [35]. Object-related CAD models are available in a highly-detailed geometric form since they are necessary attributes in manufacturing. Real-time rendering framework allows the use of these CAD models of compatible formats within a simulation. Taking advantage of this feature of the 3D rendering space and a virtual camera, the dataset capturing process can be simulated [38].

In addition to the benefits associated with synthetic data generation, it also exhibits notable challenges in terms of domain dissimilarity when compared to real-image-based test data. On the other hand, the synthetic data generated from CAD models come with the problem of being inherently different to the domain of the real images. Although there are various domain adaptation techniques in the literature such as synthetic-to-real domain adaptation in [39], which aim to modify synthetic data. There are also works, which present how photo-realistic rendering alone cannot reduce the domain gap and therefore attempt to resolve the issue by domain randomization [40,41]. Furthermore, investigating the combined effect of synthetic and real data as a hybrid dataset, without altering its nature, presents an intriguing avenue for exploration. In scenarios where there is an amount of limited real data, this approach could help leverage it, with synthetic data alongside.

3 Methodology

The synthetic data generation approaches described in this section could address the data limitations hindering the successful development of small OD models, by generating class specific balanced datasets. A photo-realistic synthetic dataset may not be necessary for some scenarios, where the generated synthetic datasets are mixed along with the available real dataset.

To validate this approach, an industrial use case of a manual assembly scenario has been taken into consideration. In the assembly process, a worker performs a circuit bread-board assembly of a previously designed configuration, using the specific electronics components, while wearing Augmented Reality (AR) glasses. The camera sensor in the AR glasses can capture the scene consisting of various objects available within the worker's field-of-view and feeds it into an OD model to perform object recognition. Based on the obtained results, a worker assistance-system in AR will assist the worker by providing instructions and visual augmentations. The worker assistance and AR methods will not be discussed further, as they are outside the scope of this paper.

After identifying the distribution of each target class in the original dataset, synthetic data generation will be used to balance the classes. With this approach, we are able to create a proportionally balanced dataset, with more instances of small objects in the combined dataset. Later, the SOTA OD models were trained in combinations of experiments and the results were evaluated.

3.1 Data Distribution

The data driven nature of DL/ML models make the data distribution a significant factor in the project's success. The focus of this paper is the detection of small objects in the manual assembly scenario, where a correct identification is crucial in the worker assistance context. As the first step of this work, the original distribution of the data was identified. There were *Five* objects in consideration in the target dataset, out of which the *Three* objects (LED, Resister & Button) fit the criteria of a small object i.e. below 32×32 pixels. The other *Two* objects (Buzzer, Arduino) are between 32×32 pixels and 64×64 pixels and are thus considered as medium-sized objects.

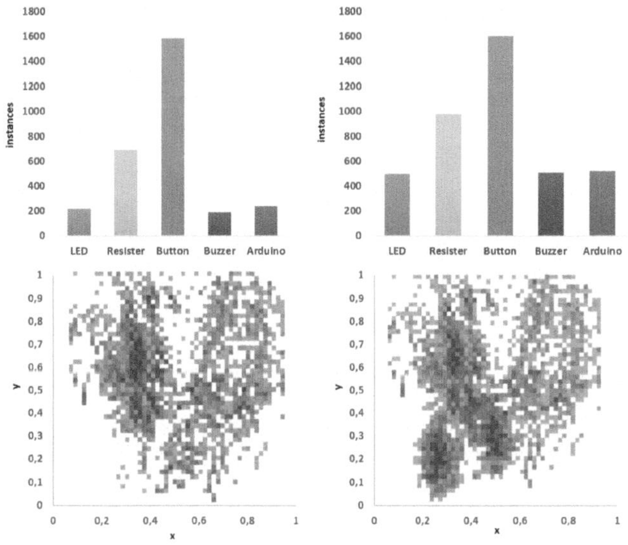

Fig. 2. Data Distribution of Initial real dataset (left) to the combined dataset DS-3 (right). The target classes have been proportionally balanced, giving more significance for the less occurring classes.

After identifying the distribution of the target objects, a proportionally balanced dataset was created using synthetic data generation. The Fig. 2 shows the initial and final distribution of the dataset (original dataset on left and the combined dataset DS-3 on the right. Details on DS-3 are shown in Table 1).

3.2 Dataset Generation

In order to create a balanced dataset, synthetic data generation method was utilized. The Table 1 lists the datasets and their combination of real and synthetic data instances used for this work. A total of *Five* datasets were generated combining the real and synthetic data, and different OD models were trained on those datasets with various combinations of hyperparameters. The datasets are later referenced using the acronyms given in this Table 1.

The datasets were generated by image capturing in a simulated scene using a virtual camera through a python script. Initially, the CAD models are imported in the simulation scene in a compatible format. On imparting the 3D models with materials and textures with minimal rendering. The dataset is simple and consists of objects in assembled and disassembled states. The object backgrounds and scales in the images are randomly set within a predefined scale range of 0.5 to 1 m from virtual camera. The illumination and object viewpoints were varied to create randomness. We used white lights and yellow lights with random illumination in the virtual scene. We do not add variable viewpoints or object occlusions in the dataset. In our use case, the objects are perceived from a limited

Table 1. The list of datasets and their division of real and synthetic data instances used. The datasets are later referenced using the acronyms given in this table.

Exp. No	Dataset Name	Real Data	Synthetic Data	Total Size of Training Data
1	DS-1	300	0	300
2	DS-2	300	100 (33% more)	400
3	DS-3	300	150 (50% more)	450
4	DS-4	300	300 (100% more)	600
5	DS-5	-	300	300

number of viewpoints. Hence, it's a simple dataset that could be generated using the available CAD models of industrial parts. A sample of synthetic image from the dataset is shown in Fig. 3.

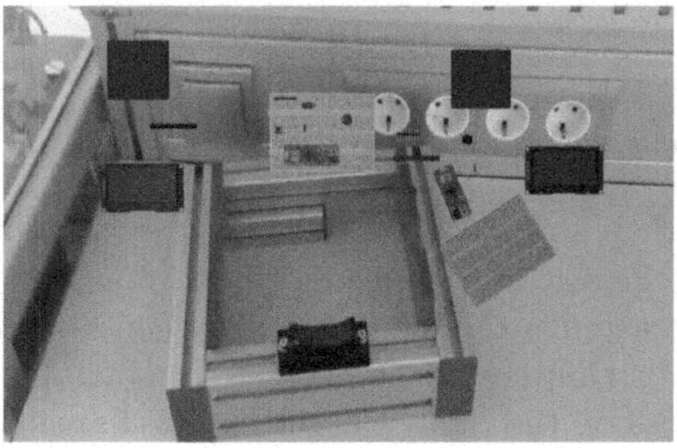

Fig. 3. A sample synthetic image from the dataset generated using the 3D rendering framework.

Figure 4 shows a sample original image from the dataset and Fig. 3 shows a synthetically generated image. As it is observable, the generated data is rather simple and mainly focuses in balancing the target object instances, rather than creating a photo-realistic scene.

3.3 Objected Detection Models

After generating the datasets and their combinations, the next step involved training and comparing SOTA OD models. For the experiments, three models were chosen based on their popularity and ease of implementation. Among them, two models were from the YOLO family, developed on the PyTorch framework, while the third model was SSD based on the TensorFlow framework. These selected models are known for their effectiveness in small OD use cases, faster inference times, and remarkable performance.

Fig. 4. A sample of real image data from the Assembly Scenario, taken from the live-stream video of AR device.

YOLOv5. YOLO (You Only Look Once) is a family of OD algorithms, which performs detection in an image through a single forward pass, unlike other algorithms like Fast-RCNN or Faster-RCNN using two stages. YOLO was the first OD algorithm to combine the procedure of predicting bounding box with class label in an end to end differentiable network [6]. YOLOv5 is open source and original version of the code is written in python with PyTorch framework. YOLO models use a Non-Maxima Suppression (NMS) as a post-processing step to obtain the final bounding box for each detected object. This technique filters out redundant bounding boxes based on their overlap values and ensures that only the most relevant and accurate predictions remain in the final results [6]. YOLOv5 offers a range of architectures, each tailored to specific application requirements and datasets. The YOLOv5s (small), YOLOv5m (medium), YOLOv5l (large), and YOLOv5x (Extra Large) architectures exhibit increasing complexity in their design. As the architecture complexity grows, so does the accuracy of the OD results. However, this improvement in accuracy comes at the expense of reduced model speed during OD. Therefore, the choice of architecture depends on striking the right balance between accuracy and real-time OD speed, considering the specific needs of the application and datasets at hand.

YOLOv7. Wang et al. in [42], introduced YOLOv7 as an extension of the YOLO series of real-time OD models. It introduces various models tailored for different GPU environments, including edge GPU, normal GPU, and cloud GPU. Examples of these models include YOLOv7-tiny, YOLOv7, and YOLOv7-W6. Additionally, YOLOv7 incorporates model scaling techniques to cater to diverse service requirements, resulting in the development of models such as YOLOv7-X, YOLOv7-E6, YOLOv7-D6, and YOLOv7-E6E. They apply stack scaling to the neck component and employ the suggested compound scaling technique to increase the depth and width of the entire model, resulting in YOLOv7-X. On the other hand, for YOLOv7-W6, they utilize the newly introduced compound scaling method to derive YOLOv7-E6 and YOLOv7-D6.

YOLOv7 in OD offers notable improvements in speed and accuracy compared to previous models. The speed ranges from 5 to 160 frames per second (FPS), enabling real-time applications. In terms of accuracy, YOLOv7 achieves the highest average precision (AP) of 56.8% among all real-time object detectors operating at 30 FPS or higher [42].

SSD. SSD (Single Shot MultiBox Detector) is a prominent OD model that considers bounding box prediction as a regression problem. It starts by selecting the anchor box with the highest intersection over union (IoU) with the ground truth bounding box and gradually refines the prediction by minimizing the loss between the predicted and ground truth boxes. This iterative regression process allows SSD to achieve precise localization of objects in the image [12]. This model provides a straightforward procedure for eliminating duplicate predictions, while SSD's regression-based approach focuses on refining the bounding box estimate by progressively minimizing the error.

3.4 Hyperparameters

After identifying and selecting the OD models as described above, the experiments have been performed. To ensure a comprehensive comparison among the three OD algorithms, it is essential to maintain consistent hyperparameters during the training and testing phases on the identical dataset. In order to achieve this, the experiments employed the following hyperparameters:

- Training Batch size : 8
- Input Image size : 1080×1080 pixels (height x width)
- Optimizer: Adam optimizer
- No of Epochs: 30, 50 & 100 Epochs for each dataset

In order to maximize the features of the small objects, higher resolution images of size 1080×1080 pixels were selected as input to the models. Since the focus of the work to understand the behaviour of adding synthetic data points on the overall performance and to avoid over-fitting chances during the training, the higher epochs were not considered for the experiments.

The OD models are then trained on the dataset combinations and evaluated on a general test dataset consisted of real images. The model performances are evaluated on the COCO test evaluation matrix [43] and the results are reviewed and observations are noted.

4 Results

After training the three OD models on the five sets of datasets with three hyperparameter combinations, the models are tested on a common test dataset, that consists of real images from the Assembly scenario. Table 2, Table 3 and Table 4 represents the evaluation results from YOLOv5, YOLOv7 and SSD models respectively with the COCO evaluation matrix [43]. More detailed results are given in Appendix 1.A for YOLOv5, YOLOv7 and SSD models. The silent feature of this table is the Average Precision and Average Recall of Small objects

(APs & ARs) which calculates the actual Precision and Recall of the model based on the true bounding box and predicted bounding box overlap for small objects (less than 32×32 pixels). This was necessary because only by observing the mAP, a true analysis cannot be made as the models are performing well on most of the cases.

Upon comparing the models, the results from Table 2 show that YOLOv5 performs the best out of all the three models, especially for small objects. Upon comparing the datasets, the models trained on a dataset containing real data components (DS-1, DS-2, DS-3, DS-5) seems to perform better on the dataset with only synthetic data (DS-4). This is due to the synthetic data generation features, as mentioned in Sect. 3.2, where the samples created were non-photo-realistic and therefore upon testing in real dataset, the model trained only on synthetic dataset seem to perform inadequately. It was also observed that as the number of synthetic data points increases (from DS-1 to DS-3), the models tend to learn more features through the targeted class balancing technique and reaches a maximum performance in the DS-3 dataset, where the number of synthetic data in the dataset is exactly the half of real data instances. Upon further increasing the synthetic dataset (DS-4 & DS-5) the models tend to learn more features from the prominent synthetic data and therefore performed poorly upon testing in the real world. And comparing the hyperparameters, the models trained for 100 epochs seem to perform better compared to the other instances.

Table 2. Experiment Results for YOLOv5 Model for 100 Epochs. The experiment instance corresponding to DS-3 seem to give the optimum results, even comparing the DS-4 with a lot more synthetic data.

No	Dataset	mAP	AP50	APs	APm	APl	ARs	ARm	ARl
1	DS-1	0.976	0.967	0.493	0.557	0.373	0.564	0.61	0.4
2	DS-2	0.955	0.952	0.543	0.605	0.366	0.594	0.654	0.397
3	DS-3	**0.986**	**0.982**	**0.594**	**0.611**	**0.369**	**0.659**	**0.667**	**0.395**
4	DS-4	0.983	0.979	0.607	0.628	0.384	0.658	0.688	0.41
5	Ds-5	0.412	0.399	0.083	0.121	0.149	0.128	0.182	0.193

Table 3. Experiment Results for YOLOv7 Model for 100 Epochs. The DS-3 dataset clearly gives the best results compared to the other dataset combinations

No	Dataset	mAP	AP50	APs	APm	APl	ARs	ARm	ARl
1	DS-1	0.951	0.946	0.48	0.566	0.361	0.538	0.621	0.393
2	DS-2	0.949	0.939	0.499	0.588	0.379	0.552	0.635	0.412
3	DS-3	**0.976**	**0.967**	**0.512**	**0.624**	**0.38**	**0.553**	**0.689**	**0.404**
4	DS-4	0.946	0.939	0.47	0.578	0.578	0.519	0.659	0.398
5	DS-5	0.374	0.364	0.122	0.102	0.129	0.144	0.173	0.246

Table 4. Experiment Results for SSD Model. SSD Model seem to perform not exceptional on the data, but comparing the overall performance, the model performed comparatively better on DS-3

No	Dataset	mAP	AP50	APs	APm	APl	ARs	ARm	ARl
1	DS-1	0.826	0.765	0.204	0.329	0.314	0.265	0.411	0.347
2	DS-2	0.813	0.752	0.147	0.301	0.315	0.217	0.368	0.352
3	DS-3	**0.850**	**0.742**	**0.190**	**0.302**	**0.329**	**0.265**	**0.379**	**0.355**
4	DS-4	0.775	0.720	0.155	0.264	0.305	0.227	0.322	0.329
5	DS-5	0.198	0.142	0.088	0.101	0.123	0.092	0.132	0.169

Comparing the recent advancements in the YOLO model families, the YOLOv7 model also performs comparatively well in detecting small objects in this work. The results of this model for different experiments which are presented in Table 3, clearly show that by increasing the number of synthetic data samples, can improve the accuracy of the model to detect small objects as well as medium and large objects, but there is a limitation for that and increasing these samples too much can affect the results negatively. The results show that the highest performance occurs for DS-3, when the number of synthetic data is equal to half of real data samples.

The results of SSD model which are presented in Table 4 show that while this model can not achieve the best results compared to the other YOLO models, using and combining an amount of synthetic data with the real data samples to train the model can affect the performance and improve the accuracy of the model. The table shows that the results of training the SSD model with only synthetic data are very weak because of the difference between the synthetic dataset and the real object images, especially considering the background. On the other hand, when the amount of synthetic data samples is half of the real dataset, the performance of the SSD model is the highest (same as the YOLOv7 model), which means combining one-third of synthetic data with two-third of real dataset can affect and improve the performance of the SSD model to achieve the best result.

5 Discussion

The task of annotating the small object in a dataset is tedious. Using synthetic data generation tools or simulation frameworks for automatic annotations is an ideal approach to created labelled data points for a supervised ML application and to improve the model performance. Combining the class balanced data generation techniques with such simulation frameworks approaches, even without implementing photo-realistic rending, better performing models can be easily generated without much efforts. For certain industrial use cases these techniques can be helpful in achieving the business goals.

In a nutshell, adding synthetic data with class balancing technique can improve the performance especially for small objects, even by synthetically generating additional data points build using simple open source CAD models. The results show an increase of up to 11.4% in precision on small objects (APs) to the base model trained only on real data set (DS-1) and best combination ratio of real and synthetic dataset (DS-3) for 100 epochs. Similar percentage increase in precision and recall can be observed for models trained on 30 and 50 epochs as well. Through targeted class balancing it is possible to improve the overall dataset and to get better performances.

6 Conclusion

In certain scenarios where real-world datasets are being limited and biased towards prominent objects, present challenges in detecting less prominent and less significant objects, such as small-sized objects. However, by modifying the dataset through *targeted class balancing techniques*, leveraging synthetic data generation, appears to improve the performance of data-driven OD models. The results from this work show that even with low quality synthetic dataset, an increase of 10% in the mAP for small objects can be achieved using this class balancing technique. As the generated data was not photo-realistic compared to the real world data, adding more synthetic images seem to degrade the model's performance. In comparison with three state-of-the-art OD Models: SSD, YOLOv5 & YOLOv7; the YOLOv5 model seem to perform better in this particular use case.

Although this study gives a clear comparison of state-of-the-art OD models, it has potential for many future works. By extending the quality of the generated images to more photo-realistic and bring closer to the real world could potentially improve the performance a lot. But the data diversity challenges in the real world and the corresponding model performance and comparing it to the costs of data generation can still debatable.

Acknowledgment. This research work is funded by the German BMBF - Bundesministerium für Bildung und Forschung (0IW19002, project InCoRAP). We would also like to thank Al-Harith Farhad for reviewing the paper.

Appendix 1.A Detailed Summary of Model-Training

See Tables 5, 6 and 7.

Table 5. Experiment Results for YOLOv5 Model, the experiment instance corresponding to DS-3 at 100 Epochs seem to give the optimum results, even comparing the DS-4 with a lot more synthetic data.

No	Epoch	Dataset	mAP	AP50	APs	APm	APl	ARs	ARm	ARl
1	30	DS-1	0.922	0.912	0.373	0.516	0.3	0.446	0.574	0.339
2	50		0.946	0.939	0.47	0.532	0.35	0.515	0.588	0.375
3	100		0.976	0.967	0.493	0.557	0.373	0.564	0.61	0.4
4	30	DS-2	0.936	0.929	0.401	0.511	0.329	0.476	0.583	0.364
5	50		0.953	0.949	0.475	0.576	0.362	0.537	0.631	0.394
6	100		0.955	0.952	0.543	0.605	0.366	0.594	0.654	0.397
7	30	DS-3	0.955	0.953	0.475	0.576	0.33	0.53	0.63	0.367
8	50		0.959	0.953	0.488	0.566	0.364	0.354	0.624	0.398
9	100		**0.986**	**0.982**	**0.594**	**0.611**	**0.369**	**0.659**	**0.667**	**0.395**
10	30	DS-4	0.966	0.959	0.449	0.594	0.346	0.484	0.661	0.375
11	50		0.977	0.968	0.577	0.613	0.382	0.627	0.672	0.402
12	100		0.983	0.979	0.607	0.628	0.384	0.658	0.688	0.41
13	30	DS-5	0.369	0.357	0.143	0.093	0.137	0.143	0.169	0.183
14	50		0.336	0.323	0.057	0.104	0.170	0.057	0.139	0.207
15	100		0.412	0.399	0.083	0.121	0.149	0.128	0.182	0.193

Table 6. Experiment Results for YOLOv7 Model. The DS-3 dataset clearly gives best results compared to the other dataset combinations

No	Epoch	Dataset	mAP	AP50	APs	APm	APl	ARs	ARm	ARl
1	30	DS-1	0.885	0.873	0.259	0.416	0.273	0.306	0.487	0.309
2	50		0.905	0.897	0.36	0.45	0.285	0.447	0.505	0.328
3	100		0.951	0.946	0.48	0.566	0.361	0.538	0.621	0.393
4	30	DS-2	0.838	0.831	0.331	0.404	0.313	0.403	0.461	0.35
5	50		0.914	0.907	0.463	0.479	0.347	0.56	0.538	0.372
6	100		0.949	0.939	0.499	0.588	0.379	0.552	0.635	0.412
7	30	DS-3	0.947	0.943	0.424	0.514	0.32	0.475	0.588	0.352
8	50		0.957	0.95	0.42	0.575	0.34	0.472	0.64	0.375
9	100		**0.976**	**0.967**	**0.512**	**0.624**	**0.38**	**0.553**	**0.689**	**0.404**
10	30	DS-4	0.906	0.902	0.305	0.538	0.319	0.349	0.618	0.354
11	50		0.928	0.922	0.372	0.51	0.322	0.434	0.616	0.359
12	100		0.946	0.939	0.47	0.578	0.578	0.519	0.659	0.398
13	30	DS-5	0.243	0.235	0.083	0.059	0.072	0.103	0.108	0.147
14	50		0.304	0.3	0.105	0.075	0.085	0.114	0.117	0.145
15	100		0.374	0.364	0.122	0.102	0.129	0.144	0.173	0.246

Table 7. Experiment Results for SSD Model. SSD Model seem to perform not exceptional on the data, but comparing the overall performance, the model performed comparatively better on DS-3

No	Epoch	Dataset	mAP	AP50	APs	APm	APl	ARs	ARm	ARl
1	30	DS-1	0.492	0.480	0.015	0.174	0.254	0.051	0.251	0.290
2	50		0.630	0.617	0.079	0.240	0.300	0.170	0.292	0.330
3	100		0.826	0.765	0.204	0.329	0.314	0.265	0.411	0.347
4	30	DS-2	0.414	0.409	0.036	0.142	0.274	0.062	0.214	0.304
5	50		0.556	0.540	0.079	0.177	0.263	0.113	0.232	0.305
6	100		0.813	0.752	0.147	0.301	0.315	0.217	0.368	0.352
7	30	DS-3	0.333	0.307	0.007	0.087	0.246	0.027	0.134	0.289
8	50		0.574	0.555	0.033	0.175	0.242	0.091	0.244	0.280
9	**100**		**0.850**	**0.742**	**0.190**	**0.302**	**0.329**	**0.265**	**0.379**	**0.355**
10	30	DS-4	0.253	0.232	0.002	0.041	0.230	0.021	0.079	0.263
11	50		0.446	0.430	0.028	0.144	0.210	0.072	0.214	0.242
12	100		0.775	0.720	0.155	0.264	0.305	0.227	0.322	0.329
13	30	DS-5	0.081	0.076	0.011	0.034	0.063	0.062	0.071	0.092
14	50		0.121	0.099	0.067	0.059	0.091	0.079	0.108	0.136
15	100		0.198	0.142	0.088	0.101	0.123	0.092	0.132	0.169

References

1. Krizhevsky, A., Sutskever, I., Hinton, G.E.: ImageNet classification with deep convolutional neural networks. In: Pereira, F., Burges, C.J.C., Bottou, L., Weinberger, K.Q. (eds.) Advances in Neural Information Processing Systems, vol. 25, Curran Associates, Inc. (2012)
2. Sermanet, P., Eigen, D., Zhang, X., Mathieu, M., Fergus, R., LeCun, Y.: OverFeat: integrated recognition, localization and detection using convolutional networks. arXiv preprint arXiv:1312.6229 (2013)
3. Girshick, R.B., Donahue, J., Darrell, T., Malik, J.: Rich feature hierarchies for accurate object detection and semantic segmentation. CoRR, abs/1311.2524 (2013)
4. Girshick, R.: Fast R-CNN. In: 2015 IEEE International Conference on Computer Vision (ICCV), pp. 1440–1448 (2015)
5. Ren, S., He, K., Girshick, R., Sun, J.: Faster R-CNN: towards real-time object detection with region proposal networks. IEEE Trans. Pattern Anal. Mach. Intell. **39**(6), 1137–1149 (2017)
6. Redmon, J., Divvala, S., Girshick, R., Farhadi, A.: You only look once: unified, real-time object detection. In: 2016 IEEE Conference on Computer Vision and Pattern Recognition (CVPR), pp. 779–788 (2016)
7. Papers with code - COCO test-dev benchmark (object detection)
8. Chen, X., et al.: Microsoft COCO captions: data collection and evaluation server. CoRR, abs/1504.00325 (2015)
9. Bochkovskiy, A., Wang, C., Liao, H.M.: YOLOv4: optimal speed and accuracy of object detection. CoRR, abs/2004.10934 (2020)

10. Borkman, S., et al.: Unity perception: generate synthetic data for computer vision. arXiv preprint arXiv:2107.04259 (2021)
11. Ren, S., He, K., Girshick, R., Sun, J.: Faster R-CNN: towards real-time object detection with region proposal networks. In: Advances in Neural Information Processing Systems, vol. 28 (2015)
12. Liu, W., et al.: SSD: single shot MultiBox detector. In: Leibe, B., Matas, J., Sebe, N., Welling, M. (eds.) ECCV 2016. LNCS, vol. 9905, pp. 21–37. Springer, Cham (2016). https://doi.org/10.1007/978-3-319-46448-0_2
13. D'Angelo, T., et al.: Deep learning-based object detection for digital inspection in the mining industry. In: 2019 18th IEEE International Conference on Machine Learning and Applications (ICMLA), pp. 633–640 (2019)
14. Usamentiaga, R., Lema, D.G., Pedrayes, O.D., Garcia, D.F.: Automated surface defect detection in metals: a comparative review of object detection and semantic segmentation using deep learning. IEEE Trans. Ind. Appl. **58**(3), 4203–4213 (2022)
15. Kang, K., et al.: T-CNN: tubelets with convolutional neural networks for object detection from videos. IEEE Trans. Circuits Syst. Video Technol. **28**(10), 2896–2907 (2018)
16. Dai, J., He, K., Sun, J.: Instance-aware semantic segmentation via multi-task network cascades. In: 2016 IEEE Conference on Computer Vision and Pattern Recognition (CVPR), pp. 3150–3158 (2016)
17. Herath, S., Harandi, M., Porikli, F.: Going deeper into action recognition: a survey. Image Vis. Comput. **60**, 4–21 (2017). Regularization Techniques for High-Dimensional Data Analysis
18. Menikdiwela, M., Nguyen, C., Li, H., Shaw, M.: CNN-based small object detection and visualization with feature activation mapping. In: 2017 International Conference on Image and Vision Computing New Zealand (IVCNZ), pp. 1–5 (2017)
19. Benjumea, A., Teeti, I., Cuzzolin, F., Bradley, A.: YOLO-Z: improving small object detection in YOLOv5 for autonomous vehicles (2023)
20. Zhang, W., Wang, S., Thachan, S., Chen, J., Qian, Y.: Deconv R-CNN for small object detection on remote sensing images. In: IGARSS 2018 - 2018 IEEE International Geoscience and Remote Sensing Symposium, pp. 2483–2486 (2018)
21. Zhou, X., et al.: Intelligent small object detection for digital twin in smart manufacturing with industrial cyber-physical systems. IEEE Trans. Industr. Inf. **18**(2), 1377–1386 (2022)
22. Zheng, X., Zheng, S., Kong, Y., Chen, J.: Recent advances in surface defect inspection of industrial products using deep learning techniques. Int. J. Adv. Manuf. Technol. **113**, 03 (2021)
23. de la Rosa, F.L., Gómez-Sirvent, J.L., Sánchez-Reolid, R., Morales, R., Fernández-Caballero, A.: Geometric transformation-based data augmentation on defect classification of segmented images of semiconductor materials using a ResNet50 convolutional neural network. Expert Syst. Appl. **206**, 117731 (2022)
24. Wang, J., Lee, S.: Data augmentation methods applying grayscale images for convolutional neural networks in machine vision. Appl. Sci. **11**(15) (2021)
25. Martins, D.H., et al.: Hybrid data augmentation method for combined failure recognition in rotating machines. J. Intell. Manuf., 1–19 (2022)
26. Salamon, J., Bello, J.P.: Deep convolutional neural networks and data augmentation for environmental sound classification. IEEE Signal Process. Lett. **24**(3), 279–283 (2017)
27. Emam, K., Mosquera, L., Hoptroff, R.: Chapter 1: Introducing Synthetic Data Generation. O'Reilly Media, Inc. (2020)

28. Figueira, A., Vaz, B.: Survey on synthetic data generation, evaluation methods and GANs. Mathematics **10**(15), 2733 (2022)
29. Andersen, P.-A., Aune, T., Hagen, D.: Development of a novel object detection system based on synthetic data generated from unreal game engine. Appl. Sci. **12**, 08 (2022)
30. Kingma, D.P., Welling, M.: An introduction to variational autoencoders. Found. Trends® Mach. Learn. **12**(4), 307–392 (2019)
31. Ho, J., Jain, A., Abbeel, P.: Denoising diffusion probabilistic models. In: Proceedings of the 34th International Conference on Neural Information Processing Systems, NIPS 2020. Curran Associates Inc., Red Hook, NY, USA (2020)
32. Goodfellow, I., et al.: Generative adversarial nets. In: Ghahramani, Z., Welling, M., Cortes, C., Lawrence, N., Weinberger, K. (eds.) Advances in Neural Information Processing Systems, vol. 27, Curran Associates, Inc. (2014)
33. Karras, T., et al.: Alias-free generative adversarial networks (2021)
34. Sagues-Tanco, R., Benages-Pardo, L., Lopez-Nicolas, G., Llorente, S.: Fast synthetic dataset for kitchen object segmentation in deep learning. IEEE Access **8**, 220496–220506 (2020)
35. Rajpura, P.S., Bojinov, H., Hegde, R.S.: Object detection using deep CNNs trained on synthetic images. arXiv preprint arXiv:1706.06782 (2017)
36. Kiefer, B., Ott, D., Zell, A.: Leveraging synthetic data in object detection on unmanned aerial vehicles. In: 2022 26th International Conference on Pattern Recognition (ICPR), pp. 3564–3571 (2022)
37. Josifovski, J., Kerzel, M., Pregizer, C., Posniak, L., Wermter, S.: Object detection and pose estimation based on convolutional neural networks trained with synthetic data. In: 2018 IEEE/RSJ International Conference on Intelligent Robots and Systems (IROS), pp. 6269–6276 (2018)
38. Planche, B., et al.: DepthSynth: real-time realistic synthetic data generation from cad models for 2.5 D recognition. In: 2017 International Conference on 3D Vision (3DV), pp. 1–10. IEEE (2017)
39. Nikolenko, S.I.: Synthetic-to-real domain adaptation and refinement. In: Synthetic Data for Deep Learning. SOIA, vol. 174, pp. 235–268. Springer, Cham (2021). https://doi.org/10.1007/978-3-030-75178-4_10
40. Tremblay, J., et al.: Training deep networks with synthetic data: bridging the reality gap by domain randomization. In: Proceedings of the IEEE Conference on Computer Vision and Pattern Recognition Workshops, pp. 969–977 (2018)
41. Man, K., Chahl, J.: A review of synthetic image data and its use in computer vision. J. Imaging **8**(11) (2022)
42. Wang, C.-Y., Bochkovskiy, A., Liao, H.-Y.M.: YOLOv7: trainable bag-of-freebies sets new state-of-the-art for real-time object detectors. In: Proceedings of the IEEE/CVF Conference on Computer Vision and Pattern Recognition, pp. 7464–7475 (2023)
43. Lin, T.-Y., et al.: Microsoft COCO: common objects in context. In: Fleet, D., Pajdla, T., Schiele, B., Tuytelaars, T. (eds.) ECCV 2014. LNCS, vol. 8693, pp. 740–755. Springer, Cham (2014). https://doi.org/10.1007/978-3-319-10602-1_48

AI Enhanced Processing of Large Parts: Volume Monitoring and Control of Robotic Gluing Application

Loukas Prezas, Zoi Arkouli, George Michalos, and Sotiris Makris(✉)

Laboratory for Manufacturing Systems and Automation, University of Patras, Rion, 26504 Patras, Greece

makris@lms.mech.upatras.gr

Abstract. This paper proposes an AI-driven approach leveraging stereo cameras and fuzzy logic enhanced PID control for monitoring and optimization of adhesive deposition while compensating for disturbances during the path following process. The suggested approach is based on point cloud processing acquired by a low-cost stereo vision system. The proposed approach has been validated in a deposition of glue case study coming from the bus and coach sector. A comparative study on the performance of mesh-based and voxels-based methods for volume calculation using point clouds has also been conducted. The results demonstrate that the precision and accuracy of the voxel-based method is better and promising enough to pave the way for expanding the application of the proposed approach to other path following robot processes.

Keywords: Smart Manufacturing · Quality Control · Point Cloud Processing · Autonomous reasoning · Industrial automation · Manufacturing processes

1 Introduction

Researchers and practitioners have proven the potential of Artificial Intelligence (AI) provide a series of benefits [1] addressing challenges at all levels of a manufacturing system, i.e. strategic, operational, and detailed, as presented by [2] etc. The main aspects that AI-enhanced solutions typically address include optimizing processes and designs, enhancing customization of products [3], enabling human-robot or robot-robot collaboration [4], production resilience and flexibility [5], reducing environmental impact, and so forth which are some of the main needs in today's manufacturing.

Especially quality monitoring and control has attracted a lot of interest with a series of emerging solutions have been used [6] improving interactive inspection, assembly verification, error handling and prevention [7] as well as quality assurance and waste reduction in human robot collaborative tasks [8]. The integration of quality control mechanisms into the robot control and the closed-loop feedback systems have been investigated together with adaptive control strategies to dynamically adjust robot parameters based on real-time quality feedback [9]. AI-based techniques are the backbone of the

A. Wagner et al. (Eds.): ESAIM 2023, LNME, pp. 106–115, 2024.
https://doi.org/10.1007/978-3-031-57496-2_11

solutions as they allow to perform among others defect and anomaly detection, pattern recognition, assessment of surface finish, [10].

Focusing on the integration of quality monitoring and control into robotic systems, there is considerable interest in studying path following processes (e.g. adhesive deposition and sealing). In this type of processes, it is required to accurately track a predefined path while making necessary adjustments to the end-effector's position and manipulating process parameters to ensure quality upheld. In [11] an automated end-effector with integrated vision sensors for online quality inspection during a sealing process is presented. The vision system monitors the texture and width of the fillet using image processing techniques and the local binary pattern algorithm. Similarly, in [12] image analysis is presented for measuring the width of beads in plasma arc welding processes. In [13] a passive vision system and digital image processing techniques are used for real-time measurements of the width and height of the weld bead in gas metal arc welding processes. An adaptive orbital system based on laser vision sensor for seam tracking, rail shape calibration and adaptive process control in welding applications is presented in [14]. Authors in [15] proposed a 2-camera vision system to extract the edge features in predefined areas of interest to detect the sealing seam.

Although there are many applications employing vision-based quality inspection, there is yet progress to be made in integrated volume monitoring and control in path following applications while maintaining system affordability. Additionally, little is the work on flexible quality monitoring able to withstand dynamic environments and parts variability while ensuring the proper deposition of adhesives [15]. Hence, an approach for simultaneous adhesive volume monitoring and robot position and velocity control is presented, capable of processing parts fed in random sequence and at no mechanically constrained initial positions. The proposed solution aims to eliminate the occurrence of defects and material waste and at the same time increase the flexibility of the workstation as explained in Sect. 2. Section 3 presents the implemented system and its testing under an industrial case study. The results and conclusions of the study are presented in Sects. 4 and 5, respectively.

2 Approach

The proposed approach addresses the simultaneous adjustment of the robot's end-effector velocity and position to ensure robust volume control and path following in dynamic environments. This is realized with the integration of a multi-purpose vision system integrated with a fuzzy logic enhanced robot control module. The vision system in eye-in-hand configuration detects and recognizes the part variant, tracks the corresponding predefined toolpath and triggers corrective actions (control module) for reliable path following. Feature extraction for part pose estimation is executed for adjusting the end-effector to the part's positioning. A deep learning model based on YoloV3 algorithm, for a number of object classes that depends on the number of variants that is necessary to recognize [16] is suggested. Point cloud processing enables to extract the part's points in the 3D scene and estimate the desired trajectory. A depth filter discards all scene points except for those that belong to the part. Then, a line fitting algorithm based on the least-squares method approximates the target trajectory path as further explained in [8].

Complementarily to the part's geometric characteristics, the vision system is also responsible to track the volume of the dispensed adhesive and lead to the corresponding correction of the process parameters to prevent process failure or waste of material. With the recognition of each variant, the system is also able to retrieve the adhesive volume setpoint for deposition. Subsequently, the vision system estimates the volume of the dispensed material in order to introduce the appropriate corrections or detect adhesive discontinuities which will eventually lead to bonding failure.

In this study two approaches are investigated for the estimation of the volume of adhesive material: 1. Mesh-based and, 2. Voxel based (Fig. 1). The volume is calculated in-process for an area of radius R close to the tip of the robot's tool (control area). Thus, after the stereo image acquisition and the point cloud generation, depth-based filtering is applied to isolate the volume of interest of the adhesive. The estimation of the volume of the complete adhesive line is feasible with the point cloud reconstruction of the aggregated in-process measurements. Given the end-effector position and hand-eye calibration transformation it is efficient to reconstruct the total point cloud and compute its volume once the process is completed.

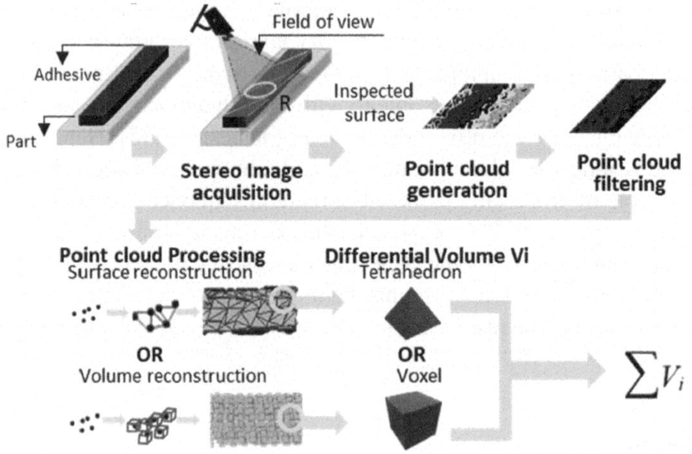

Fig. 1. Volume estimation with mesh-based and voxel-based approaches

Mesh-Based Approach: In the mesh-based approach, once the material's point cloud is defined, the final estimation is based on the Eqs. (1) (2) [17]; the volume is set as the absolute sum of signed tetrahedron volumes.

$$V_i = (-x_{i3}y_{i2}z_{i1} + x_{i2}y_{i3}z_{i1} + x_{i3}y_{i1}z_{i2} - x_{i1}y_{i3}z_{i2} - x_{i2}y_{i1}z_{i3} + x_{i1}y_{i2}z_{i3})/6 \quad (1)$$

$$V_{total} = \sum V_i \quad (2)$$

Voxel-Based Approach: In this approach an octree is used to represent the point cloud data. This enables data access as well as adaptive resolution which is important in varying densities of the points. Then, the total volume is estimated as shown in [18]:

$$V_{total=n_f} * S_{voxel}^3 \quad (3)$$

where S^3_{voxel} represents the volume of a voxel. Based on the density of the point cloud the voxel volumes are defined so the whole object is covered without overlapping.

Robot Control Approach

The adhesive volume detected close to the end-effector's tip might trigger adjustments in the robot's speed. The robot control module is responsible to translate data from the vision system into adjustments of the robot motion, which in this approach is performed iteratively and is based on a PID controller (Fig. 2.).

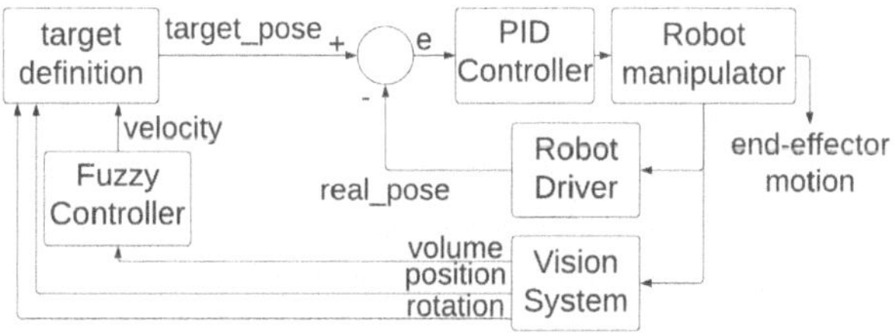

Fig. 2. Closed-loop control block diagram, b) functions.

This work focuses on three out of the six variables defining the target pose of the end-effector (Eq. 4), i.e. x, y, and rz. This is because translation along and rotation around the rest of the axes is not possible due to the part being mounted on a horizontal fixture. The target rz angle value is set based on the part's orientation for the deposition.

$$target\_pose = [x, y, z, rx, ry, rz] \quad (4)$$

The x (longitudinal axis) and y values of the *target_pose* are updated based on Eq. (5) and Eq. (6):

$$x = x + rail\_unit\_vector[0] * velocity * period \quad (5)$$

$$y = y + rail\_unit\_vector[1] * velocity * period \quad (6)$$

where *rail_unit_vector* is a vector of size 2 whose values are within [0, 1] that expresses the direction of the estimated trajectory in the x-y plane. The *period* parameter expresses the time between each target pose update. Consequently, the end-effector position and velocity will be updated based on these x, y values. The longest the target pose position is set, the fastest the robot will act to reach it.

A fuzzy logic controller is employed to estimate the end-effector's velocity target value based on volume. Three sets of volume were defined (low, average, high – triangular membership functions) and then mapped to robot velocity offsets (also with three sets defined – gaussian membership functions) as shown in Fig. 3. The membership functions' shape and parameters were finalized through experimentation; the desired performance indicates small changes in the velocity when the deposited material is around the desired volume and significant changes in other cases. As such, the shape and the interval for each function were defined to assign an appropriate velocity offset for every measured volume value, based on the defined three rules; 1. Low volume requires velocity decrease,

2. Medium volume requires no change in velocity, 3. High volume requires velocity increase. The proposed approach is more robust in noise measurements, and it can also adapt faster in comparison with an if-then-else implementation (e.g. fixed velocity offsets for specific volume values)."

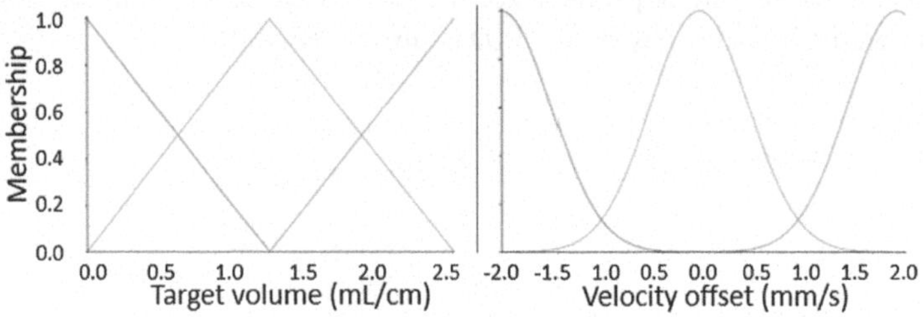

Fig. 3. Membership functions

3 Case Study

The approach has been tested and validated in a gluing case study inspired by the bus and coach sector, where structural gluing is needed to join aluminum and composite parts. The experiments have focused on the deposition of glue on aluminum rails, which can typically vary not only in their cross section and length, but also in the number of adhesive areas, and the ideal total volume that is needed for each. The developed technology aims to enhance the robotized adhesive application by providing flexibility in the objects processing and inspection by recognizing the part variants to be processed, but also ensuring adaptation to dynamic environments by detecting the parts' pose and continuously adjusting the end-effector's velocity and position to ensure that the correct amount of adhesive is deposited in the right place. Processes requiring the simultaneous control of the position of the end-effector and the volume of material being deposited can be found in many sectors such as additive manufacturing (fused deposition modeling, stereolithography, etc.), metal repairing, welding, footwear, etc.

A COMAU racer7 robot manipulator with a dispensing system, and multiple aluminum profile variants (A, B, C, D) have been used for the testing (Fig. 4), whereas ROS-based communication was used for the integration of the modules. OpenCV and PCL libraries enabled implementing customized algorithms for product and material detection, whereas a ZED2 stereo camera was used with the corresponding ROS wrapper installed to publish all data over the ROS network. NEURAL mode was activated to optimize depth sensing using machine learning techniques.

Firstly, the part type identification using the deep learning image detector, was performed and the system recognized the part. Then the fitting line approximation algorithm was enabled to estimate the desired robot trajectory on-site and provide feedback to adjust the robot motion so that the dispensing tool tip remained within the adhesive area. If the distance between the end-effector and the end point was below a set threshold (i.e., 1cm), then the gluing process was considered completed. In- process, the volume data were used for velocity control, where fuzzy logic generated a velocity target based on the observed volume and PID control to adjust the robot motion.

Variants	Volume requirements
A	0.86 ml/cm
B	0.3 ml/cm
C	1.04 ml/cm
D	0.217 ml/cm

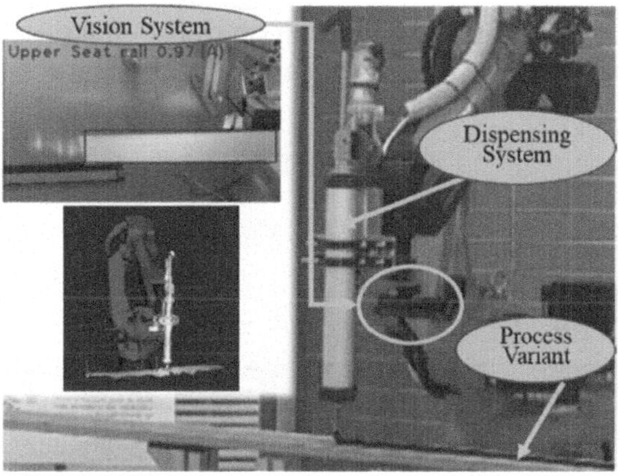

Fig. 4. Setup – Gluing process

4 Results and Discussion

Two types of tests have been performed: 1. For quality inspection, 2. Closed-loop control and the results are reported in the corresponding sections hereunder.

Quality Inspection

In this case a profile with manually pre-deposited glue volume of 220ml (the quantity was measured before use, based on mL measuring cylinder) was used. Twenty experiments were conducted where the robot moved above the profile with a steady speed to perform quality inspection. The results in Table 1 indicate that the voxel-based approach results in better accuracy and precision which refer to the volume estimation for the total amount of the deposited glue (point cloud reconstruction of the aggregated in-process measurements). In terms of runtime (time to complete the volume estimation once the point cloud is extracted for a) control areas but also for the b) complete adhesive line), the voxel-based approach is faster, which has been expected given that the mesh-based approach includes the transformation of the point cloud to mesh for each volume estimation. Figure 5 presents the volume measurements that were generated in-process for the control areas (small areas close to the end-effector) and their synthesis into providing volume measurements for the complete adhesive line using the two methods. The main purpose of Fig. 5 is to illustrate the response of the two methods in case of glue line discontinuities, as well as to compare the two methods in terms of noise. Furthermore, the point cloud of the complete adhesive line has been added on top of the plot to show the connection between the plots and the recorded 3D data. Based on the plot, the mesh-based approach does not perform reliably when there are gaps/discontinuities in the dispensed material. This is because the point cloud is considered as a united object when computing the mesh's volume, which also justifies the existence of noisy measurements compared to the other method. It has been observed that the voxel-based approach is more sensitive during specific times of the day (e.g., afternoon, >250 lx), compared to the mesh-based method It is worth noting that, the mean time between two velocity updates (including volume estimation, fuzzy controller, target definition, etc.) is 21 ms that adequate for correcting the robot's trajectory smoothly preventing adhesive line discontinuities or waste of material.

Table 1. Volume estimation methods comparison

Methods	Average Accuracy	Average Precision	Average Runtime (control area)	Average Runtime (inspection area)
Mesh-based	93,74%	± 5.73%	3.152 ms	172.3 ms
Voxel-based	95.54%	± 4.79%	0.285 ms	14.64 ms

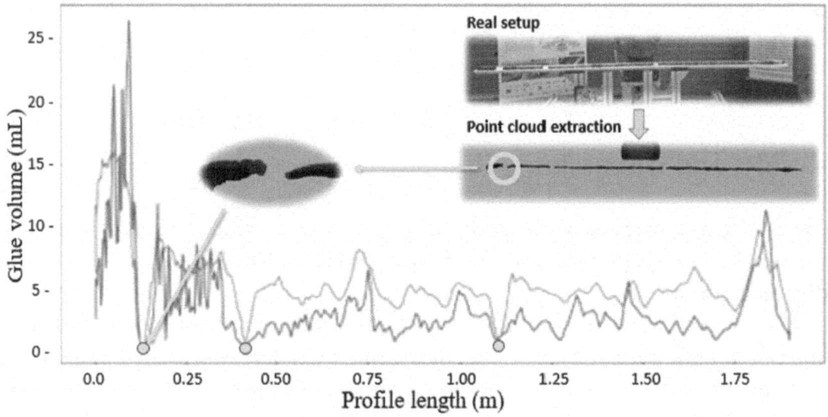

Fig. 5. Volume distribution (orange for voxel-based, blue for mesh-based)

Closed Loop Control

The voxel-based approach was selected for feeding the closed-loop control module. In this type of testing, the robot as depositing material on a profile, and multiple tests were conducted to ensure the process quality. The volume target has been 1.5 mL for 1.5 cm glue-line length, with the system's response shown in Fig. 6. The robot started with zero speed and accelerated once material was detected with corrective velocity actions held to keep the materials volume on the set target. The speed was analogue to volume (as expected), but with small delay, whereas the average glue volume was kept on 1.5 cm with the highest extremes observed being 2.69 mL and 0.98 mL. The observed oscillation of the system response is mainly caused by the extraction of the glue point cloud. The surfaces of the parts that were processed in the experiments had reflective surfaces. This in tandem with the fact that the point cloud is extracted by colour filters might lead to not detecting existing points. Moreover, it is possible that the existence of small black objects on the profiles' surface lead in detecting extra points. The volume estimation value is affected by the number and the geometry of the point cloud points which justifies the oscillations in Fig. 6.

Overall, the solution seems highly promising as it consistently achieves continuous glue lines without hills in every run, which means that the objectives of bonding safe-guarding and waste reduction were successfully achieved. However, it is noted that the observed volume peaks are quite spread out. Nevertheless, this trade-off is justified when considering the affordability of the solution compared to highly accurate systems that require substantial investments, are more complex, and demand greater maintenance. In more detail, commonplace industrial solutions for continuous volume measuring in adhesive applications are usually based on laser sensors or force sensors, which typically cost more than 5000€, whereas the sensor used costs currently approximately 450€ [19]. Furthermore, the authors will pursue optimization of the deviations by testing more algorithms for volume estimation, as well as for filtering of background and reflections, and investigating the performance of cameras with better specifications. Finally, the authors intend to perform robustness analysis and optimization under varying light conditions.

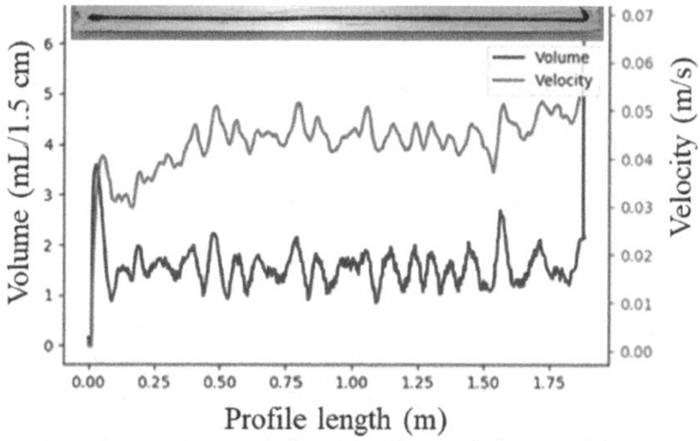

Fig. 6. System's response during closed loop control

5 Conclusions and Future Work

An approach for volume monitoring and control in path following robotic applications based on a vision system, and fuzzy logic enhanced PID control was presented. The main advantage of the presented solution is that multiple functionalities including process inspection, robot control, and part recognition are achieved cost-efficiently and with satisfactory performance. In more detail, the point cloud processing successfully detected defects, and accurately calculated the dispensed volume. Voxel-based and mesh-based approaches were compared for volume estimation with the first one performing better. Further research effort will be given to conduct experiments during different lightning conditions to present an analysis regarding the performance of the algorithm for the different values of luminosity. The voxel-based method was tested for volume-based robot control using a fuzzy logic enhanced PID controller. The results are promising in the stationary robot setup; thus, the authors aim to integrate the developed solution into a mobile robot scenario and extend the presented methodology with the necessary navigation modules. Further research will focus on optimizing the performance of the proposed algorithms for changing light conditions. Additionally, enhancing the proposed approach with flow control will be investigated as well as the integration of edge devices to optimize the system's performance, runtime and network latencies.

Acknowledgements. This research has been supported by the European project "PENELOPE Closed-loop digital pipeline for a flexible and modular manufacturing of large components" (Grant Agreement: 958303) funded by the European Commission.

References

1. Ribeiro, J., Lima, R., Eckhardt, T., Paiva, S.: Robotic process automation and artificial intelligence in industry 4.0 – a literature review. Procedia Comput. Sci. **181**, 51–58 (2021). https://doi.org/10.1016/j.procs.2021.01.104
2. Chryssolouris, G., Alexopoulos, K., Arkouli, Z.: A Perspective on Artificial Intelligence in Manufacturing (2023)
3. Gan, Z.L., Musa, S.N., Yap, H.J.: A review of the high-mix, low-volume manufacturing industry. Appl. Sci. **13**(3), 1687 (2023). https://doi.org/10.3390/app13031687
4. Makris, S.: Cooperating Robots for Flexible Manufacturing. Springer Series in Advanced Manufacturing. Springer, Cham (2021). https://doi.org/10.1007/978-3-030-51591-1
5. Kokotinis, G., Michalos, G., Arkouli, Z., Makris, S.: A behavior trees-based architecture towards operation planning in hybrid manufacturing. Int. J. Comput. Integr. Manuf., 1–26 (2023). https://doi.org/10.1080/0951192X.2023.2228254
6. Kim, H., Lin, Y., Tseng, T.-L.B.: A review on quality control in additive manufacturing. RPJ **24**(3), 645–669 (2018). https://doi.org/10.1108/RPJ-03-2017-0048
7. Antonelli, D., Stadnicka, D.: Predicting and preventing mistakes in human-robot collaborative assembly. IFAC-PapersOnLine **52**(13), 743–748 (2019). https://doi.org/10.1016/j.ifacol.2019.11.204
8. Prezas, L., Michalos, G., Arkouli, Z., Katsikarelis, A., Makris, S.: AI-enhanced vision system for dispensing process monitoring and quality control in manufacturing of large parts. Procedia CIRP **107**, 1275–1280 (2022). https://doi.org/10.1016/j.procir.2022.05.144
9. Prezas, L., Arkouli, Z., Michalos, G., Angelakis, P., Gkournelos, C., Makris, S.: A multipurpose robot perception system enabling closed-loop control for zero defect manufacturing in gluing processes of large parts. In: SSRN (2023, preprint). https://doi.org/10.2139/ssrn.4334519
10. Schmitt, J., Bönig, J., Borggräfe, T., Beitinger, G., Deuse, J.: Predictive model-based quality inspection using machine learning and edge cloud computing. Adv. Eng. Inform. **45**, 101101 (2020). https://doi.org/10.1016/j.aei.2020.101101
11. Pereira, F.R., et al.: Force and vision-based system for robotic sealing monitoring. Int. J. Adv. Manuf. Technol. **126**(1–2), 391–403 (2023). https://doi.org/10.1007/s00170-023-11110-z
12. Font Comas, T., Diao, C., Ding, J., Williams, S., Zhao, Y.: A passive imaging system for geometry measurement for the plasma arc welding process. IEEE Trans. Ind. Electron. **64**(9), 7201–7209 (2017). https://doi.org/10.1109/TIE.2017.2686349
13. Pinto-Lopera, J., Motta, J.S.T., Absi Alfaro, S.: Real-time measurement of width and height of weld beads in GMAW processes. Sensors **16**(9), 1500 (2016). https://doi.org/10.3390/s16091500

14. Silva, R.H.G.E., Galeazzi, D., Schwedersky, M.B., Mendonça, F.K., Bonamigo, A.V., Marques, C.: An adaptive orbital system based on laser vision sensor for pipeline GMAW welding. J. Braz. Soc. Mech. Sci. Eng. **43**(7), 358 (2021). https://doi.org/10.1007/s40430-021-03081-7
15. Zhou, P., et al.: Neural reactive path planning with Riemannian motion policies for robotic silicone sealing (2023)
16. Redmon, J., Farhadi, A.: YOLOv3: an incremental improvement. arXiv, 08 April 2018. http://arxiv.org/abs/1804.02767. Accessed 9 June 2023
17. Zhang, C., Chen, T.: Efficient feature extraction for 2D/3D objects in mesh representation. In: Proceedings 2001 International Conference on Image Processing (Cat. No.01CH37205), Thessaloniki, Greece, pp. 935–938. IEEE (2001). https://doi.org/10.1109/ICIP.2001.958278
18. Bienert, A., Hess, C., Maas, H.-G., Von Oheimb, G.: A voxel-based technique to estimate the volume of trees from terrestrial laser scanner data. Int. Arch. Photogramm. Remote Sens. Spatial Inf. Sci. **XL–5**, 101–106 (2014). https://doi.org/10.5194/isprsarchives-XL-5-101-2014
19. Shop the ZED Store | Stereolabs. https://store.stereolabs.com/en-eu?_ga=2.266449230.1933421517.1692435179-1502242634.1675429413&_gl=1%2Ai4wv8l%2A_ga%2AMTUwMjI0MjYzNC4xNjc1NDI5NDEz%2A_ga_LQLTWBS792%2AMTY5MjQzNTE3OC4yNC4xLjE2OTI0MzUxODIuNTYuMC4w. Accessed 19 Aug 2023

Advancing Human-Robot Interaction Using AI – A Large Language Model (LLM) Approach

Nikos Dimitropoulos, Pantelis Papalexis, George Michalos, and Sotiris Makris(✉)

Laboratory for Manufacturing Systems and Automation, University of Patras, Rion, 26504 Patras, Greece

makris@lms.mech.upatras.gr

Abstract. Human-Robot Interaction (HRI) plays a crucial role towards enabling seamless collaboration between humans and robots, leading to increased productivity, efficiency, and safety in production processes. Robots equipped with advanced sensing capabilities can work alongside humans, performing repetitive and physically demanding tasks, while humans focus on more complex and strategic aspects of production. Several concepts have been presented so far involving a variety of h/w and information exchange means, like gestures, physical and virtual buttons and voice commands. The main drawback identified that limits the fluidity of the interaction is that the commands are pre-programmed and have to be memorized. This paper presents a framework to advance HRI in both execution and programming phases, through the use of AI, exploring large language models (LLMs), allowing humans to communicate in natural language. A reference use case is presented to validate the effectiveness of this approach.

Keywords: human-robot interaction · large language models · manufacturing · human augmentation

1 Introduction

Flexible production systems not only help to serve a variety of final products but facilitate productivity. Flexibility can be achieved with the use of human operators, augmented by smart production entities, like collaborative robots [1]. Over the years, numerous concepts and approaches have been proposed towards seamless human robot interaction [2]. These encompass a diverse range of hardware and information exchange methods, including gestures, physical and virtual buttons and voice commands. Each of these modalities offers unique advantages and presents opportunities for seamless collaboration. Gestures enable intuitive and non-verbal communication through natural body movements. Physical and virtual buttons provide tangible interfaces that can be easily accessed and manipulated. Voice commands, on the other hand, offer a hands-free and efficient means of communication [3].

Moreover, efforts are being made to incorporate context-awareness into HRI systems. By equipping robots with contextual understanding capabilities, they can consider situational factors, environmental cues, and user preferences to enhance the quality of

A. Wagner et al. (Eds.): ESAIM 2023, LNME, pp. 116–125, 2024.
https://doi.org/10.1007/978-3-031-57496-2_12

interaction [4]. This allows robots to adapt their behaviour and responses based on the specific needs and preferences of individual users or the collaborative environment.

To advance human-robot interaction in a generic way though, deeper changes have to be taken into account, considering the flexibility of the system. Predefined assignment of tasks and strict robot programming are approaches that shall be omitted [5]. Instead, skills-based programming [6], coupled with teaching by demonstration, revolutionizes the way programming is taught and learned. Operators can directly teach new skills to the robots, on the spot, without programming knowledge. Usual approaches involve teaching by demonstration using manual guidance [7], AR graphics and gestures [8].

While these HRI and programming approaches do show promise, there is a recognized drawback that hampers the fluidity of interaction: the reliance on dedicated actions which require the operators to have their hands occupied. This is solved by voice commands, which on the other side consists of pre-programmed commands that need to be memorized. This limitation poses challenges to users, requiring them to learn and remember specific command sets for each robot or system they interact with.

During the last year, it has been a dramatic increase in the research, development and application of AI in a variety of fields. In this sense, utilization of large language models (LLMs), like ChatGPT, unlocked new potentials for the research community, as they provide the opportunity to develop natural language applications supporting a variety of languages for education purposes [9], health [10], and even human-robot collaboration [11], although trust of such systems is yet to be evaluated [12].

This paper presents an approach to advance HRI through the use of AI, exploring LLMs, allowing humans to communicate in natural language. Specifically, a framework is presented involving the capturing of user's input using voice commands in natural language, processing and translation to robot commands published through ROS. The novelty of this paper lays in the fact that non-expert persons can interact with collaborative robots in a natural way, without the need to memorize specific keywords, patterns or procedures, both in execution and programming phases. The use of AI through the LLMs in a pioneering manner elevated accessibility to roles necessitating industrial expertise. Additionally, its implementation holds the potential to restrict deployment durations, mitigate the effects of an aging workforce, amplify productivity, and establish an appealing work environment for the younger generation. The framework was applied in a use case from the white goods industrial sector.

2 Approach

The proposed framework is presented in Fig. 1. Natural language commands are captured by the operator and then translated to the corresponding robot commands. The approach bifurcates into two pivotal phases: the Programming Phase and the Execution Phase, both of which exist within the domain of a digital twin. During the Programming Phase, the operator can teach the robot precise target points, enable/disable the gripper on demand and store the individual points. Moreover, flexibility or strictness of robot actions order can be configured. Regarding the Execution Phase, the robot program initially taught can be used as a reference point, allowing the operator to ask the robot to deviate from it, either in task sequencing or in way of execution (e.g. alter the robot pose). Each execution cycle can serve as a new entry to robot training/programming, adapting the robot to operator needs/preferences. This iterative phase-by-phase interaction fosters a continuous loop of learning and enhancement, encapsulating the architecture's adaptability to dynamic operational requirements.

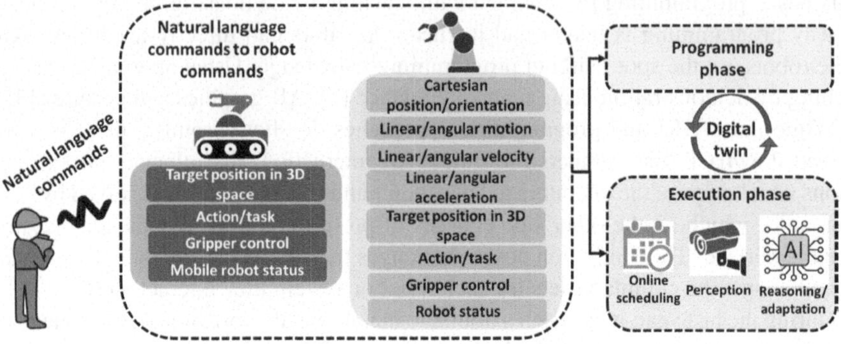

Fig. 1. Overview of the proposed framework.

In a more advanced state, the objective is to program and train a robot through natural language commands. On one hand, the training can be implemented through human-robot interaction in two phases: i) programming phase, where the operator strictly programs the robot, ii) execution phase, where the robot can continuously learn new skills through the HRI, adapting to operator needs and preferences. On the other hand, programming a robot could be implemented from another robot that is already trained. The main concept behind the proposed approaches concerns the human to robot and robot to robot programming without the need of actual programming.

2.1 Robotic Arm Commands

Initially, on robotic arm's manipulation several voice/textual commands are given as an input to execute or freeze a movement. An approach enables the robot to navigate to a designated cartesian position within the workspace having the orientation defined. Another movement approach will be for the robot to move relative to its current position by providing displacement linear values in terms of x, y, and z e.g. "move 'x' centimeters diagonally", enabling operators to precisely control and adjust the arm's position. Furthermore, some more general and subjective instructions could be provided e.g. "move (a little more) up/down/etc.". Also, robotic arm could be moved precisely towards a target by specifying the desired angle displacement, direction angular velocity and acceleration. Moreover, predefined target positions in space can be provided e.g. "go to the green workbench/ default position". In terms of advanced and intuitive HRI, another proposed aspect includes the ability to provide commands based on any coordinating system e.g. operator's one.

At a more sophisticated level, the real possibilities extend beyond elementary movement commands. Specifically, the ultimate objective extends to accommodating higher-level commands and more intricate functionalities, enabling the robots to interpret and execute sophisticated instructions. By integrating technologies such as computer vision collaborating with artificial intelligence and LLMs, the aim is to empower the framework with the ability to comprehend and respond to complex commands that involves the recognition of the cable length, material and thickness, the evaluation of connectors' placement, the type of screws, the recognition of screwing location, type of electronic components and connectors etc., rendering to a high-level user experience. Finally, another objective lays in the local usage of an LLM model that can be used on an autonomous system.

Lastly, some other actions that concern the activation or deactivation of the robot, the start or stop of a motion and the repetition of robot's movement can be provided.

2.2 Mobile Robot Commands

In a similar way, LLMs have the potential to significantly enhance the HRI capabilities on mobile robots' field where operators can employ natural language inputs to issue a variety of commands and instructions. These commands include essential functionalities such as starting or powering on the robot, as well as shutting down or powering off. Moreover, fundamental locomotion commands such as moving forward/backward/diagonally, moving towards right/left, rotating left/right or executing a u-turn by either providing or not the displacement values can be effortlessly conveyed, as well as movement to a specific target. Additionally, similarly to robotic arm commands, by integrating visual perception and mapping technologies, it will be feasible to precisely navigate the robot by specifying the location, the task and the speed on an industrial environment e.g. "move to the inventory room and bring me the box with the M10 nuts".

3 Functionalities and Implementation

The establishment of integration and communication interfaces was carried out in accordance with the principles of ROS [13]. LLM that is utilized to process the natural language commands was the model gpt-3.5-turbo-0301 of OpenAI. The specific model was selected after the evaluation of current available models such as ada, davinci, etc. An additional pivotal factor contributing to this decision was the fact that even without fine-tuning, the model yielded comparable outcomes to the fine-tuned models. The system is deployed on a personal computer running Ubuntu 20.04 and ROS Noetic.

The architecture of the implemented system is presented in Fig. 2. The operator provides textual commands or vocal commands that are translated through Google Voice Analyzer to textual commands. Robot is always in a standby state. This was achieved by running simultaneously threads that await input from operators. When an input occurs, it is stored on the bottom of the list that consists of model's description and some example prompts. Given unstructured operators' commands are extracted from the list and corresponded to JSON-structured commands through the LLM: gpt-3.5-turbo-0301. LLM response is provided as vocal and textual output to the operator, and it is stored on the bottom of the list too. Finally, these commands are interpreted to ROS commands that are responsible for joint drivers' control.

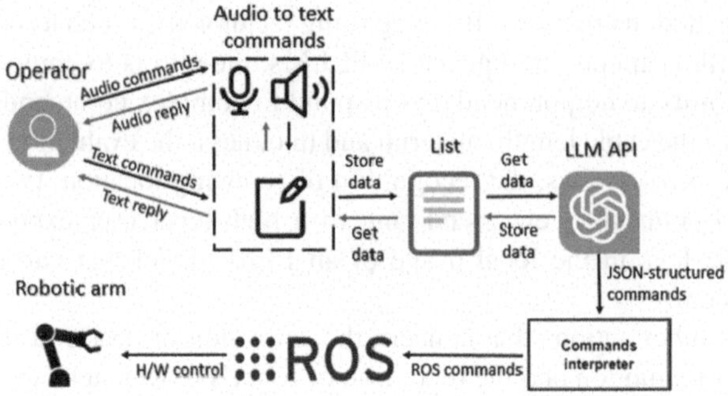

Fig. 2. LLMs method architecture for HRI.

3.1 System Functionalities

The current implementation is founded upon plain movement commands for the manipulation of a UR-10 robot that in future implementation will be a part of more complex and higher level commands. The JSON-structured commands that have been used are indicated below:

- *Up/ Down/ Front/ Back/ Left/ Right*
- *Up/ Down/ Front/ Back/ Left/ Right for specific centimeters*
- *Turn left / Turn right*
- *Turn left / Turn right for specific degrees*
- *Default, WorkbenchA, WorkbenchB* (predefined target positions)
- *Stop, Start, Pause and Resume*

The commands "*Up/ Down/ Front/ Back/ Left/ Right*" are implemented through the set_joint_value_target() of ROS. The values that are given to each of the joints are the minimum or the maximum possible values that could be provided. The commands "*Up/ Down/ Front/ Back/ Left/ Right for specific centimeters*" are implemented through the set_pose_target(). They also take as input the desired displacement and the robot moves based on the predefined coordinating system. As far as the rotation to the right and left, the value of the base joint, that is given through set_joint_value_target(), is equal to 360° or −360° respectively. In the case of rotation to specific degrees, the values are provided by the user. Additionally, predefined positions have been provided to move to a specific target point in 3D space. Corresponding commands were determined for the motion of the mobile robot, added the *"Mobile"* in front of the previous JSON-structured commands. Finally, commands to pause/resume motion are provided.

The natural language commands that have been given to the robotic arm correspond to simplified, short and accurate JSON-structured commands. These are converted to ROS commands so that the robot can be manipulated. Examples of these commands are indicated to the Table 1. The values on the set_joint_value_target() are provided on rad and on set_pose_target() on meters while the J0, J1, etc. refer to the angles of joints and x, y, z to cartesian values.

Table 1. Examples of natural language commands and the corresponding commands.

Example of verbal commands	JSON-structured commands	ROS commands
Please, move up	Up	set_joint_value_target (J0, −1.57, 0, −1.57, J4, J5)
Move front for 20cm	Front 0.2	set_pose_target (x, y + 0.2, z)
Turn right for 180°	Turn right 180	set_joint_value_target (−1.57, J2, J3, J4, J5, J6)
Come to my workbench	Workbench A	set_joint_value_target (1.6, −1.6, −2, −2.7, J5, J6)
Move the mobile robot front for 30 cm	Mobile front 0.3	set_position_target (x, y + 0.3, z)
Stop immediately	Stop	stop ()

3.2 LLM Node Programming

The LLM node consists of two classes and the main() function. The first class concerns the motion planning of the robot while the second deals with the LLM's utilization. The motion planner class contains the functions that are generated based on the packages of MoveIt planning framework. Each movement function controls the joint state, the orientation, and the position of the end effector. These functions are called on the main() function when the appropriate JSON-structured command is called.

The other class of the node concerns the prompting of LLM. Specifically, each operator's instruction is converted by using the LLM to one of the JSON-structured commands that are referred on Sect. 3.1. The LLM was already trained; thus by providing some input examples, the output completions that generates are precise and accurate. The provided input included the system's description, a number of prompts and their completions. In our case the system's description was: "Describe strictly the input with one of the JSON-structured commands that provided" while an example of prompts/completion couple is: "prompt: Move front for 90 cm-> completion: Front 0.9". While exploring available OpenAI models, fine-tuning was not available for the utilized model. As much as it concerns the available models, it was proved to be less dependable, demanding substantial resources for training while yielding comparable results.

On the main() function, ChatCompletion request of the OpenAI API was utilized, converting unstructured natural language to JSON-structured commands. Each of these commands activates the appropriate movement function. Finally, it was observed that the LLM model did not have memory capabilities. So, in order to be more interactive and acquire the capability of tracking a conversation, the LLM was provided with all previous request and responses, allowing for memorization of past commands.

4 Case Study

On this section, a reference use case derived from the white goods sector will be presented. The reference scenario focuses on assembly of induction kitchen hobs, involving two actors, namely a human operator and a low payload collaborative robot (UR-10). Both actors can work at the same time on the same workpiece. The operator loads the coils into a feeding system, locating the terminals of the cables in a fixed position. The robot using a specially designed end-effector along with a smart multi-sensor system will confirm that the coil and connectors are correctly loaded. Then it proceeds with the assembly sequence, ensuring highly reliable screwing of the connectors to the terminals. The assembly takes place at a dedicated workbench, while a secondary workbench nearby serves the raw material and tools storage.

A dedicated simulation was set up using Gazebo to replicate the reference workstation and apply the developed solution. Vocals instructions were given through a simple headset's attached microphone, and they were converted to textual instructions that were provided through the node terminal. The coordinating system of the robot, except from the z axis, is the opposite of operator's system. On the Fig. 3 an example is shown which presents a group of command from the categories analyzed on chapter 3.

Fig. 3. Examples of natural language commands.

The robotic arm is always on a stand-by mode waiting for the operator's instructions. Instructions could be provided even during the execution of a robot's movement. At the specific occasion, on condition that the command is not the "Stop", a new path planning will take place and the robot will start the movement towards the new target. Except for movement commands, while a human incorporates with a robot, it is crucial and constructive for the robot to move to specific target points into the 3D space. In current implementation four target points are provided and the operator asks for the movement execution towards them. Another important aspect is that the LLM can effectively keep track of a conversation. On the current implementation as it is referred on the Sect. 3, the previous conversations are stored into a list that is provided to the LLM on each of its requests. As a result, the operator could ask for instruction repetition. In addition to the opportunity for movement repetition, another useful feature that was provided by making the track list of previous conversations, concerns the capability of "re-train" the

LLM. So, LLM can generate different JSON-structured commands than the ones before, when the same input is provided.

An experimental procedure was set up to preliminary evaluate the effectiveness of the developed solution, aiming to gather feedback that will help to pave the roadmap for additional developments and enhancements. Five subjects (4 males, 1 female), with age variability of 23–37 years old participated in the experiments, testing the developed functionalities and assessed them in terms of user experience. The participants were experienced developers and users with previous knowledge of AR tools while most of them are owners of a commercial voice assistant. As a general remark, based on their feedback, LLM's responses exhibited remarkable accuracy, even in cases where the sentences were not well structured or contained typographical errors. Furthermore, it was highlighted that the generated responses emulated human conversation patterns while the conversational flow and the degree of interaction was elevated. Notably, participants expressed that the LLM's responses facilitated a heightened ease in maintaining a coherent conversational track.

At the end of the experimentation, the operators filled in questionnaires to capture their subjective feedback to assess the user-friendliness of the system, regarding the system usability, accessing aspects as attractiveness, clarity, efficiency, satisfaction and novelty. Attractiveness concerns emotional engagement and overall experience. Clarity is essential too, dealing with the understandability of the system's interface. Efficiency plays a significant role, as evaluating the system's workflow and task completion without unnecessary steps. Additionally, assessing users' satisfaction reveals their level of contentment, comfort, and fulfillment, reflecting the system's impact on their overall experience. Lastly, the element of novelty is related to the innovative features or functionalities of the system's compared to existing solutions.

The system usability score was 86.4 which indicated usability rating as "excellent" according to the benchmark (Fig. 4). The subjects evaluated the system Attractiveness, Efficiency, Satisfaction and Novelty as "excellent", and only Clarity score was rated as "good".

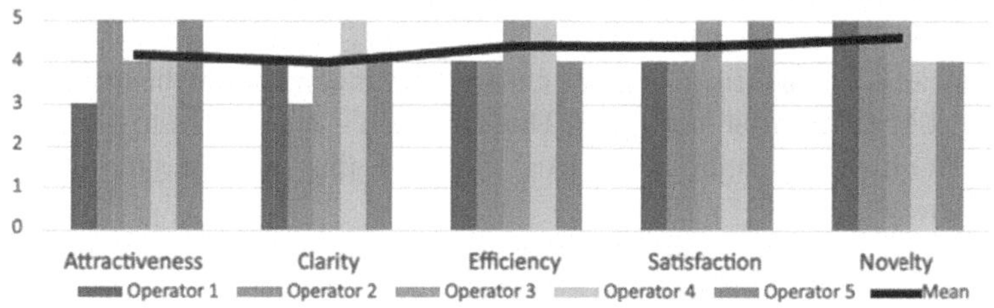

Fig. 4. User Experience ratings.

5 Conclusions and Future Work

This paper presented a framework for using LLMs to facilitate interaction in human robot collaborative applications. Employing AI through LLMs in a pioneering manner enhances access to industrial expertise roles, potentially reducing deployment time, addressing workforce aging, boosting productivity, and creating an attractive environment for younger professionals. The framework consists of functionalities that support both the execution as well as programming phases, allowing teaching of new robot programs and skills, supporting both stationary and mobile robots.

Regarding the aspect concerning the LLM, its significance lies in the imperative necessity for the development of a more integrated solution. Current implementation rests upon the OpenAI API. However, inherent limitations exist within this framework. Instances of network connectivity issues or instances of excessive utilization can result in unresponsive LLM response. Furthermore, the scope of available tokens accessible during the model prompting phase remains constrained. A notable constraint concerns the inherent resource demands of LLMs. Owing to their expansive datasets and algorithm complexity, necessitate substantial memory and computational resources for efficient operation within a local system. These constraints, coupled with the financial implications associated with each individual API request, have collectively contributed to a more challenging implementation process.

In the present implementation, the constraints did not pose an obstacle thus the framework's design and testing were executed within a simulated environment for one robotic arm in testing mode. The quantum of requests made remained notably distant from the prescribed limitations, as the API key was exclusively accessed by a single user at any given time. The LLM model that employed was the gpt-3.5-turbo-0301, selected due to its prevailing potency and extensive feature set. Furthermore, the absence of the necessity for exhaustive fine-tuning rendered it the most expedient and high-value alternative.

The main aspect highlighted by participants in their feedback evaluation pertains to the responses generated by the LLM. This observation is underscored by the transformation of LLM responses into JSON-structured commands, which, upon output, can resemble encrypted messages. To enhance the HRI in future iterations, prospective strategies include the modification of the current LLM's behavior or the integration of an additional LLM to function in parallel. This approach is anticipated to yield responses distinct from those originating solely from the LLM, thus contributing to an enriched HRI experience.

Future work will aim towards the implementation of the remaining functionalities of the framework related to human to robot and robot to robot programming for a complete cycle, followed by more perspective and higher-level commands and functionalities related to control of mobile robots. The overarching goal is to enable robots to interpret and execute intricate commands, achieved through integration with technologies like computer vision, artificial intelligence, and LLMs. This empowers the framework to understand complex instructions involving diverse attributes such as cable details, connector evaluation, screw types, and electronic component types, etc.

The solution will be demonstrated in physical setups involving use cases from white goods, automotive, aerospace and additive manufacturing cases in context of CONVERGING EU Project. Lastly, the authors plan to release this framework as open source.

Acknowledgements. This research has been supported by the EU project "CONVERGING Social industrial collaborative environments integrating AI, Big Data and Robotics for smart manufacturing". This project has received funding from the European Union's Horizon Europe research and innovation programme under grant agreement No 101058521.

References

1. Chryssolouris, G.: Manufacturing Systems: Theory and Practice, 2nd edn. Springer, New York (2006). https://doi.org/10.1007/0-387-28431-1
2. Michalos, G., Karagiannis, P., Dimitropoulos, N., Andronas, D., Makris, S.: Human robot collaboration in industrial environments. In: The 21st Century Industrial Robot: When Tools Become Collaborators, pp. 17–39 (2022)
3. Dimitropoulos, N., Togias, T., Michalos, G., Makris, S.: Operator support in human–robot collaborative environments using AI enhanced wearable devices. Procedia CIRP **97**, 464–469 (2021)
4. Rodriguez-Guerra, D., Sorrosal, G., Cabanes, I., Calleja, C.: Human-robot interaction review: challenges and solutions for modern industrial environments. IEEE Access **9**, 108557–108578 (2021)
5. Dimitropoulos, N., Michalos, G., Makris, S.: An outlook on future hybrid assembly systems-the Sherlock approach. Procedia CIRP **97**, 441–446 (2021)
6. Herrero, H., Abou Moughlbay, A., Outón, J.L., Sallé, D., de Ipiña, K.L.: Skill based robot programming: assembly, vision and workspace monitoring skill interaction. Neurocomputing **255**, 61–70 (2017)
7. Massa, D., Callegari, M., Cristalli, C.: Manual guidance for industrial robot programming. Ind. Rob. Int. J. **42**(5), 457–465 (2015)
8. Tsarouchi, P., Athanasatos, A., Makris, S., Chatzigeorgiou, X., Chryssolouris, G.: High level robot programming using body and hand gestures. Procedia CIRP **55**, 1–5 (2016)
9. Kasneci, E., et al.: ChatGPT for good? On opportunities and challenges of large language models for education. Learn. Individ. Differ. **103**, 102274 (2023)
10. Biswas, S.S.: Role of chat GPT in public health. Ann. Biomed. Eng. **51**(5), 868–869 (2023)
11. Koubaa, A.: ROSGPT: Next-Generation Human-Robot Interaction with ChatGPT and ROS. Preprints.org (2023)
12. Ye, Y., You, H., Du, J.: Improved trust in human-robot collaboration with ChatGPT. IEEE Access **11**, 55748–55754 (2023)
13. Pyo, Y., Cho, H., Jung, L., Lim, D.: ROS Robot Programming; ROBOTIS, Seoul, Korea (2017)

Cognitive Exoskeletons: Harnessing AI for Enhanced Wearable Robotics in Shipbuilding

Alexandros Kanakis, Konstantinos Katsampiris-Salgado, Natalia Zacharaki, Nikos Dimitropoulos, and Sotiris Makris(✉)

Laboratory for Manufacturing Systems and Automation, University of Patras, 26504 Rion Patras, Greece
makris@lms.mech.upatras.gr

Abstract. In recent years, manufacturing has been undergoing major changes in search of more flexible and intelligent solutions, powered by Industry 4.0. While automation is what drove previous progress in industry, not every process can be automated, therefore, there is still a need for humans to carry out more complex tasks. Thus, in order to increase productivity and efficiency, while maintaining flexibility, robots can assist operators in their tasks rather than replace them. This paper proposes techniques for achieving seamless Human – Robot Collaboration (HRC), with wearable robots equipped with cognitive abilities that make them aware of their environments and the tasks the operator is performing. The main focus is on semi-active exoskeletons with adjustable torque and how they can be made autonomous. Their awareness is powered by computer vision techniques, mainly using an RGB-D camera and can automatically adjust the level of assistance they provide to the operator based on what they perceive, rather than having the operator control them directly. Consequently, they do not only improve the workload of the operator, but this assistance is offered without requiring specific actions from the user. This solution mainly focuses on the shipbuilding industry and the various tasks the workers have to perform in which exoskeletons can aid them. Finally, a welding use case is presented demonstrating the various functionalities.

Keywords: Machine vision · Human action perception · Gesture recognition · Wearable robotics · Exoskeletons

1 Introduction

The growing need for reconfigurable production systems and flexible manufacturing [1] has led to the integration of Human-Robot Collaboration (HRC) into industrial settings [2]. This approach combines human versatility with robotic precision and repeatability, especially valuable in complex sectors like shipbuilding, where customization makes mass production ineffective [3]. Thus, enhancing operator productivity through advanced tools becomes crucial.

However, the human element often leads to musculoskeletal disorders due to repetitive, heavy tasks [4]. Exoskeletons, which is a distinct category of HRC, rather than being standalone entities, they act as physical extensions of the human body augmenting

A. Wagner et al. (Eds.): ESAIM 2023, LNME, pp. 126–135, 2024.
https://doi.org/10.1007/978-3-031-57496-2_13

his capabilities [5]. By providing limb support and reducing fatigue, exoskeletons prove especially beneficial in industries where tasks require heavy lifting or static limb positioning over long durations. However, an optimal level of support is needed, which is not overly restrictive and does not inhibit operator's range of motion [6]. These devices need optimal support levels that do not inhibit the users' motion [7].

Control methodologies for exoskeletons range from electromyography (EMG) [8] to traditional controls like knobs [6], buttons [9], and mobile applications [10]. Despite their utility, methods like EMG present challenges due to daily fluctuating, subject-dependent parameters [11].

Human action and activity recognition is a broad research field that focuses on identifying and classifying human behaviours, movements, and activities from different sensors pipelines. The choice of sensor systems is often influenced by specific application requirements. Data from multiple sensor sources can be combined, providing accurate data capturing. Wearable sensors like IMUs located at human joints, smart devices like smartphones and smartwatches that capture acceleration or velocity data, physiological sensors (e.g., EEG, ECG, EMG) assessing operators' physiological states, and vision-based systems (which might utilize RGB, depth, or thermal data) are all examples of these sensor systems [12].

Utilizing different sensors simultaneously often involves the application of multi-sensor information fusion algorithms. Techniques such as statistical and probabilistic methods (e.g., Weighted average, Kalman filters), classical machine learning algorithms (e.g., Support Vector Machines, Hidden Markov Models, Decision Trees), deep learning techniques (e.g., Convolutional Neural Networks, Long Short-Term Memory Networks), and transfer learning strategies (e.g., Domain Adaptation, Reinforcement Transfer Learning) are applied in these scenarios [12].

For human action and activity recognition, once the data is processed and possibly fused, various methods, primarily based on artificial intelligence, are employed. These methods range from supervised and unsupervised learning, metaheuristics, reinforcement learning, deep learning, to ensemble learning [13]. Additionally, there are also examples of probabilistic approaches in the literature [14].

This paper presents an innovative control approach in wearable robotics. Considering the range of tasks in the shipbuilding sector, that usually contain overhead operations (e.g., overhead welding, sandblasting and water jet blasting) thus causing ergonomic strains on the operator's shoulders, a Human Action Perception (HAP) module is proposed by the authors to optimally control an upper limb exoskeleton device leveraging AI and machine vision. HAP, utilizes a head mounted camera and computer vision to recognize user actions and detect the tools that are used, thus aiming to autonomously adjust exoskeleton assistance level (without human manual inputs) based on the requirements of each specific task. Factors such as the operator's focus, hand usage, and overall intention are crucial in tailoring this assistance. The novelty of the proposed system lays upon the extension of human limits, improving ergonomics, which is a major consideration in tasks with high physical effort.

2 Approach

The main objective is the development of an AI methodology, utilizing computer vision techniques to recognize the activity of an operator while performing specific tasks. It focuses on a specific use case, to calculate an appropriate Level of Assistance (LoA) that will be provided by a semi-active exoskeleton.

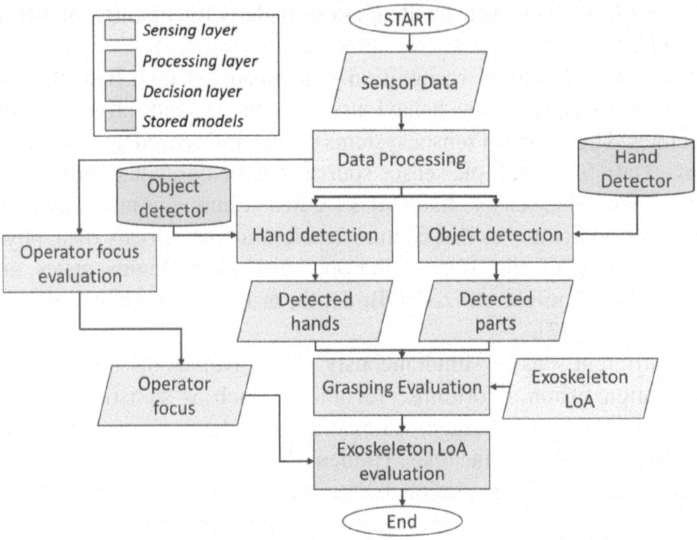

Fig. 1. HAP flowchart

To this end, a Human Action Perception (HAP) module was developed combining several features that enable autonomous exoskeleton control, without human input for LoA adjustment. The main features are (i) Hand recognition and location tracking, (ii) Equipment recognition and tracking, (iii) Part grasping identification, (iv) Activity recognition and intention prediction and (v) Focus detection – head orientation. All these features are combined in order to determine a final Level of Assistance (LoA) that the operator is going to receive from the exoskeleton.

In the specific case, the exoskeleton is a shoulder assistive device that provides antigravitational support to the user's arms for activities requiring permanent or repetitive shoulder flexion for long time periods. The exoskeleton aids the operator's shoulders through two actuation boxes equipped with a spring-loaded mechanism that produces the assistive torque. Additionally, a servomotor adjusts the springs' pre-tension level. The assistive torque is embedded within the spring-based system and aims to relief the shoulder from the gravitational torque of the human arm. An energy-efficient adjustment mechanism, paired with a servomotor, facilitates the automatic alteration of the spring's tension [15]. Thus, different LoAs that the HAP set, correspond to change of the tension of the spring.

The HAP module consists of three layers. The first one is the sensing layer, and it includes all the sensory network that is utilized to receive data such RGB images,

IMU acquisitions, barometric pressure values etc. The sensor data are then shared to the processing layer, where their analysis is performed, including filtering and reshaping. During this stage, the data are categorized based on their type to allow for input into the appropriate algorithm or deep learning network for decision-making. Afterwards, the results are inserted in a decision-making algorithm, to decide the required level of assistance of the exoskeleton for the specific task the human performs; this comprises the output layer of HAP. The described approach is depicted in Fig. 1.

3 Implementation

3.1 Hardware Architecture

Hardware

Prior to presenting the software system, it is important to describe the hardware architecture, as sensor and control layouts are aspects that affect the software architecture. Firstly, a multimodal sensor was placed and fixed on the front side of the safety helmet that the operator wears, thus allowing the HAP module to perceive data similar to what operator perceives. The sensor used is the ZED2 from Stereolabs and contains multiple sensors streams, such as RGB, depth, IMU, barometer and magnetometer data, that are all utilized for human action perception a. The data pipelines are collected by a NVIDIA Jetson Xavier (NJX) which is an AI-enabled edge computing platform. Its compatibility, its processing capabilities for neural network applications along with its low consumption led to choosing it.

The NJX device is connected to a mini router via ethernet to enable quick data sharing and access to local network which includes the NJX and the exoskeletons processing unit, which are connected to the router via Wi-Fi. All modules are supplied by a high-capacity power bank, and they are placed in the operator's back. The hardware interconnection is presented in Fig. 2.

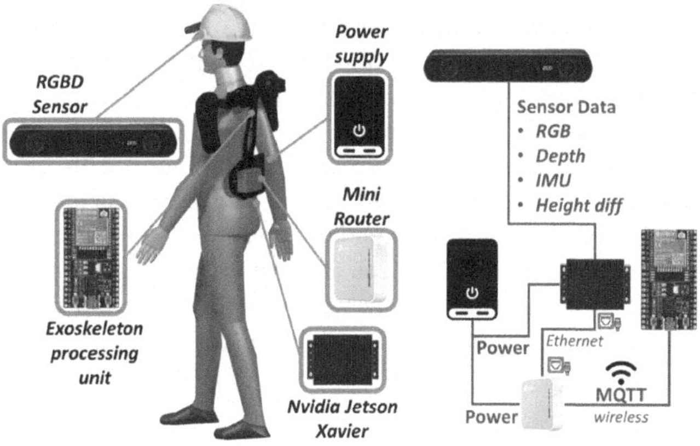

Fig. 2. HAP hardware architecture

Communication
The protocol used to share the LoA commands to the exoskeleton is MQTT, which is a lightweight messaging protocol designed for efficient communication between devices. It is widely used in Internet of Things (IoT) and machine-to-machine (M2M) applications. It uses a subscriber – publisher messaging transport. In this case, the exoskeleton's control module subscribes to several topics and the HAP module publishes the LoA as an integer to two of these.

3.2 Object Detection

Object Detection Model and Position Detection
To detect the tools the operator is using, a Machine Learning (ML) object detection model was trained and deployed. The model should overcome four main common challenges such as high-accuracy redundant detection, false positive elimination, occluded tool detection, and recognition of tools in various poses.

Different object detection models were considered: Faster Region-based Convolutional Neural Network (Faster RCNN), Single Shot Detection (SSD), and You Only Look Once (YOLO). These models, among the most prevalent in the field of Machine Vision (MV), differ in terms of detection accuracy and speed.

Consideration was also given to the performance capabilities of NJX, particularly its balance of performance, portability, and low power consumption. In this context, real-time detection speed is a critical metric for this use case. In the conducted experiments, YOLO outperformed the other models, in detection accuracy, leading to its selection.

Detecting the objects' positions is crucial for determining their use. If a tool is held, the LoA is calculated based on its weight. Distance from the camera is measured using the ZED's depth sensor. The object's location is determined within its bounding box by matching points to the point cloud. The closest point signifies the object's position. However, if hands obstruct this area, a different point outside the overlap is selected.

Dataset Creation
Creating a training dataset for the object detection model requires consideration of lighting changes, background variations, and diversity of tools in the deployment environment. While access to the operator's tools allows for capturing representative images, it is not always possible, necessitating synthetic datasets, as suggested by [16].

Our pipeline for synthetic dataset creation includes: (i) developing 3D tool models with appropriate textures, (ii) simulating a realistic environment with variable lighting, backgrounds, and virtual cameras in a 3D software like Blender in our case, (iii) using scripts to automate object posing and lighting adjustments within the scene, including manual modifications. The software can automatically annotate the renders consistently, reducing dataset creation time. However, the time required for precise 3D model recreation and scene development, requiring skilled users, is a consideration.

3.3 Human Action Perception

Hand Tracking

The method proposed utilizes hand recognition to track tool location and usage. The Mediapipe library in Python, through a machine learning model, identifies 21 hand landmarks, aiding in hand positioning and distance calculation from the camera. When both depth and pixel proximity distances fall below a set threshold, the hand is inferred to be near the tool. Further checks confirm whether the hand is grasping the tool. Hand landmark pixel coordinates cand define a bounding box, aiding in camera-tool distance calculation even when hands obscure tools. If hand and tool bounding boxes intersect, an alternate point within the tool's box is chosen for point cloud matching.

In contrast, the second method uses ArUco markers, detectable by the OpenCV Python library, to ascertain hand coordinates, particularly beneficial when operators wear gloves that hinder hand details detection by Mediapipe. The pixel coordinates of the ArUco marker's centre are calculated, compared with the tool's bounding box coordinates, and used to estimate the camera distance. Proximity to the tool implies grasp. This method doesn't verify the grasping gesture or calculate bounding boxes due to a lack of hand landmark data. An illustration is shown in Fig. 3.

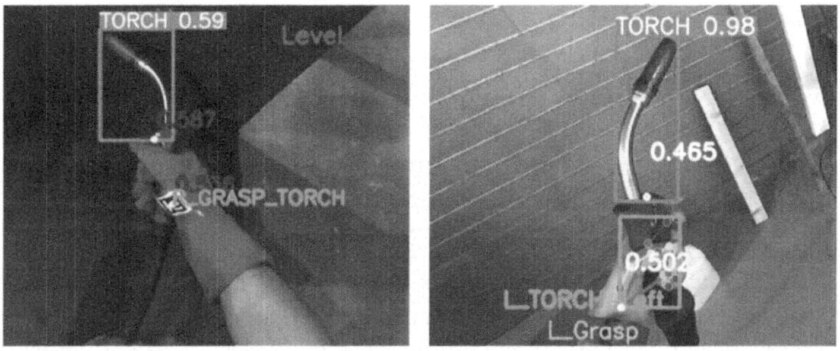

Fig. 3. ArUco based hand detection (left) and Mediapipe based hand detection (right)

The ArUco method can detect hands even when obscured by an object, while Mediapipe struggles to detect a gloved hand holding a larger tool from a certain viewpoint. In summary, Mediapipe, favoured for grasping gesture detection, and the ArUco marker method, suitable for scenarios involving gloved operators.

Head Orientation Detection

Another crucial aspect in calculating the final Level of Assistance (LoA) determined by the HAP module is the tool's position that is held by the operator. To identify when a tool is held overhead, a method was devised to detect operator head tilt using the ZED camera's IMU sensor, which is helmet-mounted, hence reflecting head orientation. The camera's orientation is obtained by calling functions from the ZED SDK that return a quaternion, allowing roll, pitch, and yaw calculation. The pitch factor is used to discern when the camera tilts upwards or downwards, assuming that the operator is looking at the tool while handling it. After LoA calculation based on other factors, a final check determines whether the operator is looking up, down, or maintaining a neutral position.

Human Focus Detection

A method for detecting human focus was developed to distinguish between when the operator is performing a task and when they are simply moving around the environment. It is assumed that during a task, the operator will maintain focus and stop moving their head, allowing for their intention to be inferred. Focus detection can also act as a signal to wait for a specific trigger. Pixel intensity values monitoring within time is not the best option, as different operations like welding and the crowded and structed industrial environment can cause changes without actually corresponding to focus change. As mentioned in the head orientation section, the camera's pose can be calculated. Then, if the quaternion values or roll, pitch, yaw values do not change beyond a certain threshold for several seconds, it is assumed that the operator is focused on a task, as the camera and consequently their head remain relatively still.

3.4 Level of Assistance Adjustment

The individual technologies used to decide upon what the operator is doing or intends to do have been described. These features need to work in tandem to calculate the appropriate LoA. As stated in the beginning, the LoA is related to the spring's tension and there are five predefined values, ranging from 1 to 5 (higher means more support is provided by the exoskeleton). The proper support from the exoskeleton is important to combine both ergonomic benefits by minimizing the weight burden of the operator's hand along with the part that he/she holds, without though making human motions stiff thus blocking the natural motion of the arm.

In most cases, the LoA is calculated as follows: the perception pipeline identifies the location of operator's hands and different objects. Depending on the hand tracking method, an evaluation for whether the operator holds a part is performed. Every object detected, has a predefined corresponding load to compensate. The final LoA is determined upon the weight of the grasped part, depending on the task that is being detected, e.g., welding or waterjet cleaning and the height of the performed action. Focus detection is mandatory to act as a blocking mechanism to prevent LoA changes during the execution of the tasks by the operators. It is worth noting that user preferences can vary. Therefore, before initiating the process, users can adjust the LoA values. For instance, for weights exceeding 2 kg, the LoA value can be increased by either 1 or 2 units, depending on individual preferences.

4 Case Study

To validate the accuracy and demonstrate the developed method, a task where the operator performs welding was chosen. There are four main tools that the operator would be using: i) an electrode, ii) a torch and iii) two types of clamps.

Two experiments were carried out, one where the operator is not wearing gloves and one where the operator carries out a welding task to completion wearing the appropriate gloves. The mapping of hands to parts is performed using the methodology described in Sect. 3. The operator carries out the task as normal. First, they secure the clamp, next they grab the welding torch or the electrode and finally they turn on the machine.

Figure 4 shows the detected hands, which part they are grasping and whether it is the left or right hand, including the distance of the objects from the camera. During welding, a bright light is produced, and the tools will stop being detected. Thus, if the center of the image is suddenly illuminated, this results in the pixel intensity values peaking in that region. So, if the operator is holding the welding torch and the mean average value of this region rises significantly, it can be inferred that welding is performed.

Before the level of assistance is finalized, a check is performed to determine if the operator is focused with their head tilted upwards, before the task initialization. For example, if the task is performed overhead and the level of assistance needs to be increased. After task initialization any LoA change is prevented by the algorithm to avoid stymie task execution. Both experiments successfully mapped hands to parts, with the Mediapipe method additionally detecting a grasping gesture for more accurate identification of held objects, an edge over the ArUco marker method. Task initialization detection, from welding light monitoring was consistent.

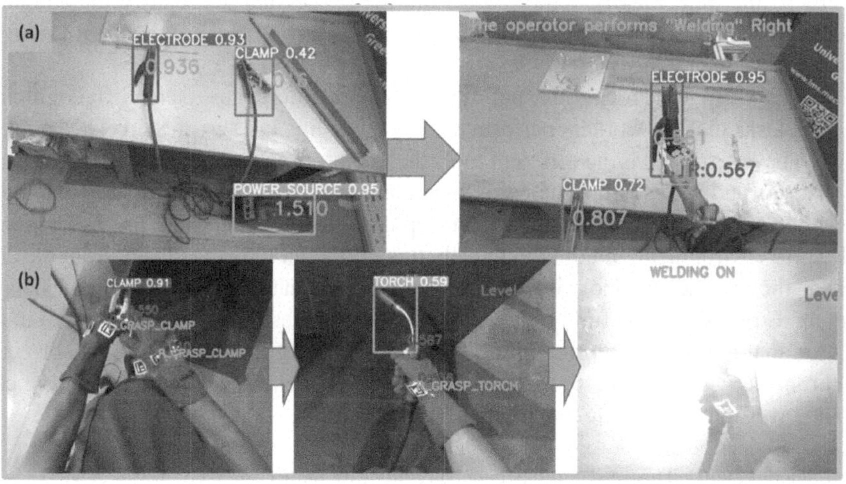

Fig. 4. In (a) the clamping positioning and in (b) the welding operation are depicted.

5 Conclusions, Discussion and Future Work

This paper presents the implementation of a Human Action Perception module that recognizes the various tasks an operator performs. It focuses on the maritime sector and its usefulness for determining the LoA of a semi-active exoskeleton support the workers wearing it. The benefit of this approach is being able to automatically change the parameters of the exoskeleton without requiring actions from the user. Also, tests were performed to determine the best object detection model for this use case and a method for creating synthetic datasets was presented. This innovation significantly enhances task recognition and adaptive support in near real-time, particularly demonstrated in welding scenarios. Its importance lies in augmenting human efficiency, safety and ergonomics, a critical concern in physically demanding jobs. This advancement in exoskeleton technology, advanced computing in portable processing units, and AI illustrate a pivotal shift towards more intelligent, human-robot collaboration, underlining its potential impact in humans according to the Industry 5.0 paradigm especially for the demanding shipbuilding sector.

A few considerations regarding the practical implementation of our system ma ybe found below. Firstly, the portable computational systems that operate the AI models and the HAP algorithms currently do not support real-time processing, primarily due to their lower specifications compared to industrial computers, compromising in processing speed and efficiency. Additionally, streaming data from sensors to a central processing unit via the internet poses significant challenges. The high payload of data often leads to latency issues, impacting the system's responsiveness. Another concern is the battery life of the portable system; frequent replacements and recharging after only a few hours of operation can lead to operational delays. This is particularly problematic when restarting the processing unit. Finally, remote access to the processing unit can be challenging, especially when operators are working in confined spaces within the shipyard, a common scenario during welding operations.

Future work should focus on experimentation with different use cases, in recognizing different tasks that the workers perform as only a welding task was tested. Therefore, it needs to be determined if the different tools the operators use are available to create real life datasets and if this is not the case, to create synthetic datasets in order to train the object detection models needed. The next step is deploying and testing this module in a real-world environment, not only in the lab. It is important to get feedback from the end users on the accuracy of recognition. Moreover, it is known that there will be some delays from the LoA calculation time to when it actually changes on the exoskeleton so there are tests that are needed on the timing of communicating the LoAs, so they correctly synchronized with what the operator is doing at any given time. Cloud computing can also be investigated utilizing 5G and IoT technologies.

Acknowledgement. This research has been supported by the EU project "Mari4_YARD User-centric solutions for a flexible and modular manufacturing in small and medium-sized shipyards". This project has received funding from the European Union's Horizon 2020 research and innovation programme under grant agreement No 101006798.

References

1. Chryssolouris, G.: Manufacturing Systems: Theory and Practice. Springer Science & Business Media (2013). https://doi.org/10.1007/0-387-28431-1
2. Inkulu, A.K., Raju Bahubalendruni, M.V.A., Dara, A., SankaranarayanaSamy, K.: Challenges and opportunities in human robot collaboration context of Industry 4.0 - a state of the art review. Ind. Robot Int. J. Robot. Res. Appl. **49**(2), 226–239 (2021). https://doi.org/10.1108/IR-04-2021-0077

3. Zacharaki, N., Dimitropoulos, N., Makris, S.: Challenges in human-robot collaborative assembly in shipbuilding and ship maintenance, repair and conversion (SMRC) industry. Procedia CIRP **106**, 120–125 (2022)
4. Alexopoulos, E.C., Tanagra, D., Konstantinou, E., Burdorf, A.: Musculoskeletal disorders in shipyard industry: prevalence, health care use, and absenteeism. BMC Musculoskelet. Disord.Musculoskelet. Disord. **7**(1), 1–10 (2006)
5. Cai, M., Ji, Z., Li, Q., Luo, X.: Safety evaluation of human–robot collaboration for industrial exoskeleton. Saf. Sci.. Sci. **164**, 106142 (2023)
6. Van Engelhoven, L., Poon, N., Kazerooni, H., Barr, A., Rempel, D., Harris-Adamson, C.: Evaluation of an adjustable support shoulder exoskeleton on static and dynamic overhead tasks. In: Proceedings of the Human Factors and Ergonomics Society Annual Meeting, vol. 62, no. 1, pp. 804–808. SAGE, Los Angeles, CA (2018)
7. Gunasekara, J.M.P., Gopura, R.A.R.C., Jayawardane, T.S.S., Lalitharathne, S.W.H.M.T.D.: Control methodologies for upper limb exoskeleton robots. In: 2012 IEEE/SICE International Symposium on System Integration (SII), pp. 19–24. IEEE (2012)
8. Lenzi, T., De Rossi, S.M.M., Vitiello, N., Carrozza, M.C.: Proportional EMG control for upper-limb powered exoskeletons. In: 2011 Annual International Conference of the IEEE Engineering in Medicine and Biology Society, pp. 628–631. IEEE (2011)
9. Otten, B.M., Weidner, R., Argubi-Wollesen, A.: Evaluation of a novel active exoskeleton for tasks at or above head level. IEEE Robot. Autom. Lett. **3**(3), 2408–2415 (2018)
10. Grazi, L., Trigili, E., Proface, G., Giovacchini, F., Crea, S., Vitiello, N.: Design and experimental evaluation of a semi-passive upper-limb exoskeleton for workers with motorized tuning of assistance. IEEE Trans. Neural Syst. Rehabil. Eng.Rehabil. Eng. **28**(10), 2276–2285 (2020)
11. Fleischer, C., Hommel, G.: A human–exoskeleton interface utilizing electromyography. IEEE Trans. Rob. **24**(4), 872–882 (2008)
12. Qiu, S., et al.: Multi-sensor information fusion based on machine learning for real applications in human activity recognition: state-of-the-art and research challenges. Inf. Fusion **80**, 241–265 (2022)
13. Ariza-Colpas, P.P., et al.: Human activity recognition data analysis: history, evolutions, and new trends. Sensors **22**(9), 3401 (2022)
14. l'Yvonnet, T., De Maria, E., Moisan, S., Rigault, J.P.: Probabilistic model checking for human activity recognition in medical serious games. Sci. Comput. Program.Comput. Program. **206**, 102629 (2021)
15. Grazi, L., et al.: Kinematics-based adaptive assistance of a semi-passive upper-limb exoskeleton for workers in static and dynamic tasks. IEEE Robot. Autom. Lett. **7**(4), 8675–8682 (2022)
16. Manettas, C., Nikolakis, N., Alexopoulos, K.: Synthetic datasets for deep learning in computer-vision assisted tasks in manufacturing. Procedia CIRP **103**, 237–242 (2021)

An Approach for Bin Picking in High-Mix Low-Volume Manufacturing

Michael van Bekkum(✉), Yulia Terzieva, Nikoletta Nikolova, and Jesse van Oort

TNO, The Hague, The Netherlands
{michael.vanbekkum,yulia.terzieva,nikoletta.nikolova, jesse.vanoort}@tno.nl

Abstract. Short production cycles and customization have led to increasing demand for flexible, adaptive solutions in industrial automation. Bin picking plays a crucial role as it can provide a significant increase in productivity and efficiency in manufacturing processes. It is the ability of a robot to choose and pick an individual object from a pile of randomly arranged ones. This field has seen ample research in recent years, with a lot of solutions presenting various uses of Artificial Intelligence techniques. However the majority of the implemented machine learning methods still lack the ability to generalize sufficiently. This is especially important in the case of high-mix low-volume situations, where the objects tend to be unique and non-uniformly shaped. These situations are commonplace in e.g. 3D printing manufacturing lines, where training a single algorithm for bin picking proposes a challenge. To tackle this we propose a modular pipeline, which splits the problem into sub-questions, each of which refers to a separate component of the picking process. This enables creation of solutions, which consist of multiple approaches combined into one. We create a first experimental implementation by addressing two parts of the modular workflow - detection and selection. This combines existing state-of-the-art bin picking techniques into a single pipeline. Preliminary results show the challenges from creating modular solutions. We argue that the benefit of such a solution is that it would enable interoperable applications.

Keywords: Bin Picking · Manufacturing · Pipeline · Detection · Matching

1 Introduction

Bin picking has been a highly researched problem for more than five decades [1] and it is currently gaining attention from major corporations [2]. It is a repetitive task, which is mostly done by hand, and involves picking objects with random poses from a filled bin. This can be automated to improve efficiency in factories and remove the need for human labour in tasks such as product sorting, quality inspection lines, bin organization, etc.

A. Wagner et al. (Eds.): ESAIM 2023, LNME, pp. 136–143, 2024.
https://doi.org/10.1007/978-3-031-57496-2_14

Automated bin picking requires a multilayered solution due to its inherent complexity. It is a combination of multiple sub-problems, ranging from image segmentation and object classification, to pose and grasp estimation [3]. Additionally, there is an extra layer of robot control, requiring accurate path planning [4] and motion control for seamless task execution. Combining all these tasks has resulted in a lot of diverse high complexity state-of-the-art methods for bin-picking. Depending on the items in the bin and their rigidness, different grippers (two- or three-jaw grippers, vacuum grippers, etc.) can be used [5]. Depending on the material of the items, there could be different camera needs regarding type and placement (e.g. for transparent items colour cameras should not be used). Given the degree of complexity of bin picking, a generalized solution that keeps in mind practical problems like high complexity items and item entanglement has not yet been developed.

This is especially true for the domain of high-mix low-volume bin picking, which is an important application in 3D printing factories and the main focus of this paper. From information presented earlier, it is evident that high-mix low-volume bin picking needs to be able to deal with small numbers of a variety of objects. We propose a pipeline approach, splitting the problem into sub-components, each of which focuses on a separate step of the process. This provides a way to introduce modularity into created solutions. Furthermore, we present an experimental implementation of part of the proposed method. We combine two existing state-of-the-art methods into a single pipeline, which provides an exact grasping point for a chosen object based on provided CAD models and a RGB-D image of the bin. Such proposed pipeline requires no training and is not limited by data or pressured by training time, meaning it can work with unseen before items, given that the CAD models are available.

The rest of the paper is structured as follows - Sect. 2 outlines the state-of-the-art, Sect. 3 describes the methodology, Sect. 4 looks at our experimental implementation, Sect. 5 provides a critical view of the discussion and lastly Sect. 6 provides a short conclusion.

2 Related Work

In the field of bin picking, there are multiple types of existing solutions, which can be split into two categories - symbolic (model-based) and data-driven (model-free). Model-based bin picking makes use of the model of the target object(s), whereas data-driven bin picking mostly relies on image segmentation and neural networks [3]. An approach that uses both data-driven and knowledge-driven components can be categorized as a Hybrid AI system [6].

In the context of bin picking, the problem can be divided into two fundamental sub-questions: *"Where is the item that needs to be picked?"* and *"How do we pick that item"*. Assuming that the CAD models of the items in the bin are available, together with colour and depth data from the bin, the first question can be answered using a Point Pair Feature-Based pose estimation algorithm (PPF). This model-based type of algorithm defines the 6D pose (3D position and 3D orientation) of a given object. The first such algorithm was created by Drost et al.[7] and since then it was been improved by many researchers ([8–10]).

Simultaneously, with the evolution of deep learning models, 6D pose estimation was performed on RGB images [11,12]. Developments like the EfficientNet [13] family of algorithms and the one-shot detector architecture YOLO [14] aided in introducing data-driven solutions. Current 6D Pose estimation methods like EfficientPose [15] and DualPoseNet [16] leverage these algorithms as base architecture for their method, achieving satisfactory results. However, these methods require substantial data for training, which can be scarce in situations with high object variability.

In the case of grasp estimations, solutions are data-driven. A research group from Berkeley called BerkelyAutomation has developed Dexterity Network (Dex-Net), which includes datasets and algorithms for robot grasping tasks [17]. They combine real-world and synthetic data and employ two types of grippers for their test cases - a suction cup and a parallel jaw. Building upon their work, Kumra et al. [18] created the Generative Residual Convolutional Neural Network (GR-ConvNet), a network which generates antipodal grasp points given a colour and depth image of a scene.

Alongside pose and grasp estimation, entanglement and multi-object picking are two other relevant challenges. Current solutions ([19,20]) make use of a data-driven approach to tackle entanglement, occurring either during grasp point prediction or during robot arm movement prediction.

The work of Zhang et al. [19] is focused on disentanglement during grasp point prediction. The approach is model-free, and generates a heat map displaying potential entanglements and selects a grasping point where the probability of entanglement is low. In situations where two objects were picked simultaneously the robot arm drops the items in a separate bin, from which they are picked again. This technique has been shown to effectively improve disentanglement [19].

The approach of Moosmann et al. [21] for disentanglement is at the robot arm movement prediction stage of bin picking. In the case that only entangled objects are left in the bin, the object with the lowest degree of entanglement is selected. A depth image of the bin is cropped such that only the selected object is in the frame and this image is passed to a pre-trained neural network, which outputs a path trajectory. The initial supervised learning implementation, although promising, requires over 300 h of dataset creation and training. In later works, Moosmann et al. [20] mitigated the training issues by using reinforcement learning for training the network, decreasing the required time to less than 10 h [20,21].

In this work we present a combinatory solution with an approach towards the bin picking issues of substantial data requirements and lengthy training. We provide an example implementation of our method and furthermore, we discuss the challenges and complexities in such cases.

3 The Workflow

A bin-picking problem can be tackled in a modular way, by splitting the problem in multiple sub-questions, which need to be answered. We propose a workflow, shown in Fig. 1, which constitutes of three stages - Data Extraction (DE), Bin Processing (BP) and Motion Planning (MP). Every stage corresponds to an element, which is needed within a bin picking solution, resulting in set of questions which need to be answered:

- **DE**: What objects do we consider? *Look for CAD models of an object*
- **DE**: What does the bin look like? *Look for camera snapshot of the bin*
- **BP**: Are there entangled regions? *Look for safe areas to grasp*
- **BP**: Which object do I grasp? *Look for fitting object*
- **BP**: How do I grasp the object? *Look for exact point*
- **MP**: How do I execute the grasp? *Look for a motion profile*

The first two questions, corresponding to the first component of the Workflow are concerned with the data necessary for a solution. The next three steps account for the different bin actions, which are needed for accurate detection - entanglement, item choice and grasp detection. The last concerns the actual picking motion that the robot needs to perform. Every element can be developed individually, resulting in a plug-and-play sort of a solution. This not only fragments the problem into smaller steps, but also allows for interchangeability of the different components.

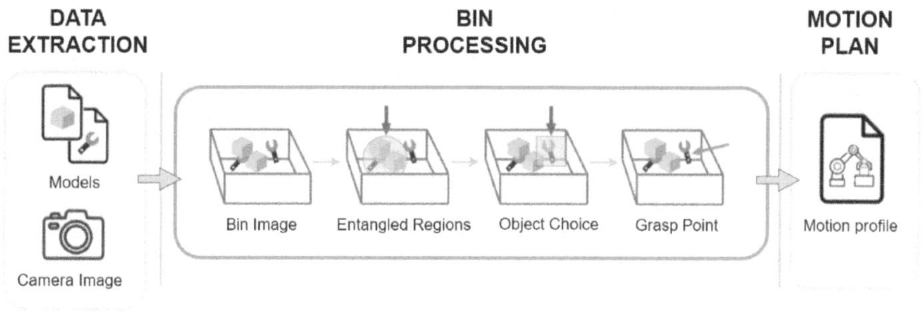

Fig. 1. Workflow structure for a proposed bin-picking solution. The image is split into three parts - (1) Most left is the Data Extraction; (2) Middle is Bin Processing, (3) Right is Motion Control

It is important to note that we make the assumption that the types of objects in the bin are known (which for the case of 3D printing is possible, as there is a model of every printed part available).

4 Experimental Implementation

To experiment with the above-described workflow, we created an example implementation focusing on the object choice and grasp generation (last two steps of the bin processing stage of the pipeline in Fig. 1). This is done by combining two existing state-of-the-art approaches, PPF [8] and GR-ConvNet [18], into a single solution process visualized in Fig. 2. We considered two main requirements when creating the solution - use state-of-the-art solutions that have an existing implementation and are published open-source. This is important as we focus not on developing the separate algorithms, but on their modularity and ability to be coupled to each other.

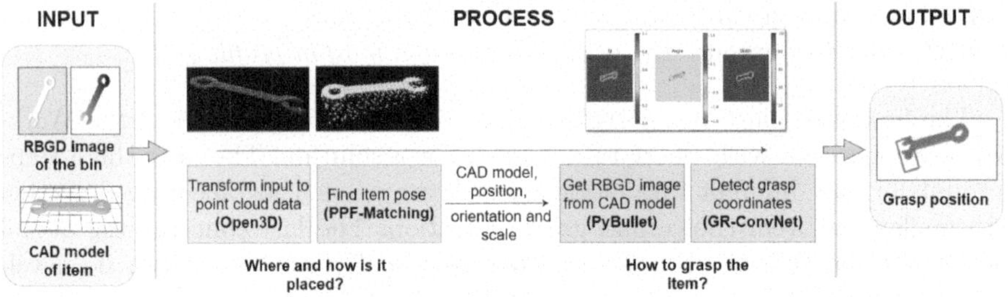

Fig. 2. Implemented workflow

First step is to provide input to the algorithm in the form of a color image of the bin, a depth image of the bin and the CAD model of the item of interest. The bin image is preprocessed into a point cloud by combining the images and removing the background.

Due to the generalization issues of data-driven approaches for pose estimation as discussed in Sect. 2, we use a Point Pair Feature (PPF) algorithm [8], which takes the point cloud of the bin and the CAD model. The algorithm outputs a 4×4 transformation matrix for every found match, and ranks them based on Hough voting. The best match is chosen as the object to be grasped.

Next, we transform the output of the PPF algorithm, using the open-source simulation tool PyBullet [22]. The orientation of the item is extracted from the transformation matrix. This is used to load the model in the simulation tool, in order to get a depth and colour image. Those images are used as input to the GR-ConvNet [18]. This outputs the coordinates of the grasp point together with the width and angle required for the gripper to execute the grasp.

To visualize how the implementation works we show in Fig. 2 what every step looks like for the scenario of a bin with a single wrench in it. The point cloud of the bin is generated and the background is removed. The PPF algorithm finds the orientation of the wrench and passes it to the grasp network. We can see that the selection of grasp is done using the three variables - quality, angle and width.

5 Discussion

Tackling the problem of high-mix low-volume bin picking in a modular way, as shown in the proposed pipeline in Fig. 1, has several advantages. It cuts the problem into smaller components, allowing for specific, specialized implementations for every component. It provides a set of fragmenting questions, aimed to assist in identifying relevant functionality. It could also aid in creating flexible interoperable solutions. The current approach, however, is presented as guidance and requires validation to be properly adjusted to reflect the needs of bin picking problems.

We aimed to begin this validation by presenting the experimental implementation. While obviously not yet in a complete state, we were able to assess and draw preliminary conclusions. Our initial analysis points towards the challenges of implementing a modular solution - while an algorithm can be very good at the task it needs to do, that does not guarantee seamless connection with other components or end-to-end optimization of the resulting pipeline.

The main insight that already emerges from combining just two state-of-the-art solutions is versatility - given the chosen algorithms, there is no need for retraining every time a new object is introduced, making the solution flexible and able to deal with new items. What we encountered, however, is that there are significant limitations to both solutions used, particularly regarding combining PPF [8] and ConvNet [18] together. Firstly, using a PPF algorithm increases the time complexity scaled with the number of different items available in the bin. Secondly, PPF removes information regarding how an item is related to others in the bin, essentially eliminating context. This could directly result in generating a grasp point that is obstructed or out of reach for the robot.

Based on those preliminary results, the first step in future would be to adapt the current experimental implementation. On one hand it is important to focus on mitigating the detected limitations. Regarding the time complexity, one approach would be to introduce an extra filtering step (based for e.g. on item frequency, entanglement etc.), which would select only relevant CAD models to match. Regarding the lost context for an item, a solution would be to introduce entanglement (Fig. 1). This could provide information regarding context to the ConvNet component, point to regions where items could be obstructed and consequently increase the probability of a successful grasp. This type of filtering could also be used to create focus zones (areas with less entanglement), which can be tackled first.

Aside from the obvious improvement of the implementation, it is important to also focus on the complete methodology. When introducing an approach, it is essential to consider its suitability for various use cases. While bin picking solutions, and by extension the presented solution, are considered to be context-independent, it should be explored whether there are factors which could create extra requirements. Such case-specific specifications could influence what part(s) of the pipeline are relevant and should be implemented. Furthermore, there is need for a validation step, where the method proposed is tested in different settings (both simulation and demo setups) to assess its suitability for real-life use-cases. This way a more robust and flexible methodology can be developed, which provides a universal approach towards high-mix low-volume bin picking problems.

6 Conclusion

In this paper we discussed the importance of using a modular approach when creating bin picking solutions to high-mix low-volume bin picking problems. In particular, we propose a three stage methodology, which partitions the general problem into sub-problems, allowing for tackling the issues one at a time. We then provided a preliminary experimental implementation using state-of-the-art approaches for two of the sub-steps - pose estimation and grasp generation. Although limited in scope, we showed that a solution for bin picking cases can be constructed by coupling existing algorithms. We also discovered that connecting stand-alone algorithms brings extra complexity, creating the need for a multi-component solution. We concluded that the chosen solution requires tuning, calibration and extension for it to be a suitable choice. Furthermore, the methodology itself needs to be thoroughly validated and adjusted to account for different use cases and their influencing requirements.

Acknowledgements. The research leading to these results has been funded by the European Commission in the H2020 research and innovation program, under Grant agreement no. 957204, the "MAS4AI" project (Multi-Agent Systems for Pervasive Artificial Intelligence for Assisting Humans in Modular Production Environments).

References

1. Buchholz, D.: Bin-picking—5 decades of research. In: Bin-Picking. SSDC, vol. 44, pp. 3–12. Springer, Cham (2016). https://doi.org/10.1007/978-3-319-26500-1_2
2. Mitash, C., et al.: Armbench: an object-centric benchmark dataset for robotic manipulation. In: ICRA 2023 (2023)
3. Cordeiro, A., Rocha, L.F., Costa, C., Costa, P., Silva, M.F.: Bin picking approaches based on deep learning techniques: a state-of-the-art survey. In: 2022 IEEE International Conference on Autonomous Robot Systems and Competitions (ICARSC), pp. 110–117. IEEE (2022)
4. Mac, T.T., Copot, C., Tran, D.T., De Keyser, R.: Heuristic approaches in robot path planning: a survey. Robot. Auton. Syst. **86**, 13–28 (2016)
5. Hernandez, J., et al.: Current designs of robotic arm grippers: a comprehensive systematic review. Robotics **12**(1), 5 (2023)
6. van Bekkum, M., de Boer, M., van Harmelen, F., Meyer-Vitali, A., ten Teije, A.: Modular design patterns for hybrid learning and reasoning systems. Appl. Intell. **51**(9), 6528–6546 (2021)
7. Drost, B., Ulrich, M., Navab, N., Ilic, S.: Model globally, match locally: efficient and robust 3D object recognition. In: 2010 IEEE Computer Society Conference on Computer Vision and Pattern Recognition, pp. 998–1005 (2010). ISSN: 1063-6919

8. Hanh, L.D., Hieu, K.T.G.: 3D matching by combining CAD model and computer vision for autonomous bin picking. Int. J. Interact. Des. Manuf. **15**(2), 239–247 (2021)
9. Cui, X., Yu, M., Wu, L., Wu, S.: A 6D pose estimation for robotic bin-picking using point-pair features with curvature (cur-PPF). Sensors **22**(5), 1805 (2022)
10. Deng, J., Qu, W., Fang, S.: A high accuracy and recall rate 6D pose estimation method using point pair features for bin-picking. In: 2022 34th Chinese Control and Decision Conference (CCDC), pp. 6056–6061 (2022). ISSN: 1948-9447
11. Xiang, Y., Schmidt, T., Narayanan, V., Fox, D.: PoseCNN: a convolutional neural network for 6D object pose estimation in cluttered scenes. arXiv:1711.00199 (2018)
12. He, Y., Huang, H., Fan, H., Chen, Q., Sun, J.: FFB6D: a full flow bidirectional fusion network for 6D pose estimation. CoRR, vol. abs/2103.02242 (2021)
13. Tan, M., Le, Q.V.: EfficientNet: rethinking model scaling for convolutional neural networks. In: Proceedings of the 36th International Conference on Machine Learning, vol. 97 of Proceedings of Machine Learning Research, pp. 6105–6114, PMLR (2019)
14. Li, C., et al.: Yolov6: a single-stage object detection framework for industrial applications. CoRR, vol. abs/2209.02976 (2022)
15. Bukschat, Y., Vetter, M.: Efficientpose: an efficient, accurate and scalable end-to-end 6D multi object pose estimation approach. CoRR, vol. abs/2011.04307 (2020)
16. Lin, J., Wei, Z., Li, Z., Xu, S., Jia, K., Li, Y.: DualposeNet: category-level 6d object pose and size estimation using dual pose network with refined learning of pose consistency. In: Proceedings of the IEEE/CVF International Conference on Computer Vision (ICCV), pp. 3560–3569 (2021)
17. Mahler, J., Matl, M., Satish, V., Danielczuk, M., DeRose, B., McKinley, S., Goldberg, K.: Learning ambidextrous robot grasping policies. Sci. Robot. **4**(26), eaau4984 (2019)
18. Kumra, S., Joshi, S., Sahin, F.: GR-ConvNet v2: a real-time multi-grasp detection network for robotic grasping. Sensors **22**(16), 6208 (2022)
19. Zhang, X., Domae, Y., Wan, W., Harada, K.: Learning to dexterously pick or separate tangled-prone objects for industrial bin picking. CoRR, vol. abs/2302.08152 (2023)
20. Moosmann, M., et al.: Separating entangled workpieces in random bin picking using deep reinforcement learning. Procedia CIRP **104**, 881–886 (2021)
21. Moosmann, M., et al.: Using deep neural networks to separate entangled workpieces in random bin picking. In: Weißgraeber, P., Heieck, F., Ackermann, C. (eds.) Advances in Automotive Production Technology – Theory and Application. A, pp. 238–246. Springer, Heidelberg (2021). https://doi.org/10.1007/978-3-662-62962-8_28
22. Coumans, E., Bai, Y.: Pybullet, a python module for physics simulation for games, robotics and machine learning (2016–2022). http://pybullet.org

Risk-Aware Task Sequencing for Human-Robot Collaboration

Alex Bonini[1], Amedeo Cesta[2], Marta Cialdea Mayer[1], Andrea Orlandini[2](✉), and Alessandro Umbrico[2]

[1] ROMA TRE University, Rome, Italy
[2] National Research Council of Italy, CNR-ISTC, Rome, Italy
andrea.orlandini@cnr.it

Abstract. Deploying collaborative robots in manufacturing presents diverse challenges. Rapid adaptability to the environment while ensuring user safety and engagement is paramount. Existing human-aware task sequencing solutions often lack explicit risk modeling and management. International standards emphasize severity, exposure, and avoidance as critical risk factors. To enhance intelligent risk awareness control, we propose integrating multiple risk factors into task sequencing models. This forms the basis for a cutting-edge planning framework-backed risk-aware task sequencing system. Our approach's evaluation across various scenarios showcases its efficacy and adaptability to diverse risk levels. Experimental results show a positive equilibrium between productivity and safety, achieving both high throughput and low operator risk.

Keywords: Human-Robot Collaboration · Artificial Intelligence · Task Planning and Scheduling

1 Introduction

The introduction of collaborative robots in manufacturing environments allows to automate challenging, hazardous, and repetitive tasks. This boosts the flexibility of manufacturing processes, although task coordination between humans and robots remains a critical issue for productivity [18]. Human-Robot Collaboration (HRC) entails challenging safety issues as collaborative environments involve tight proximity between robots and human workers raising multiple challenges [23]. It is paramount to endow collaborative robots (cobots) with the ability to quickly adapt behaviors to the actual state of the environment and keep the user safe and engaged in the interaction. Human-aware task sequencing solutions have been proposed to address task assignment and human-robot coordination (see, e.g., [14,20,22]) where risk management is often not explicitly modeled. The main risk factors highlighted in international robotic standards are severity, exposure, and avoidance (see e.g. ISO 10218, ANSI-RIA R15.06-2012). The combination of multiple factors enables a more detailed representation of risks overcoming a flat description based on a single value expressing severity and probability, which does not capture the nuances hidden in a scenario.

A. Wagner et al. (Eds.): ESAIM 2023, LNME, pp. 144–154, 2024.
https://doi.org/10.1007/978-3-031-57496-2_15

In this paper, we propose a novel risk-aware model for human-robot coordination. The model relies on widely accepted industry standards considering five flexible and general risk factors which can be applied to any HRC context. Then, We present also a software prototype planner in order to reason over the different risk factors entailed by the execution of collaborative tasks and synthesize safe and efficient collaborative task plans. Specifically, we model HRC task assignment problems as human-aware task planning problems. A task planning system then generates task plans implementing sequences of actions to accomplish predefined production objectives [4,5,14]. This entails building a flexible temporal plan that transforms the initial state of a system into a desired goal state. AI planning algorithms utilize domain knowledge and search techniques to efficiently explore the space of possible plans to identify the most feasible solution. This model informs a heuristic-based search that optimizes task assignment by considering a multi-objective perspective based on safety and efficiency. The designed heuristics allows the planner to find safe plans without compromising the efficiency of HRC productions. Its most important feature involves evaluating the risk of a situation by taking into account what the robot and the human are doing at the same time. Through simulations, we compared our planner with other strategies, collecting data on their resulting schedules. Our planner demonstrated improved task assignments, effectively balancing safety and efficiency.

2 State of the Art

In HRC, ANSI-RIA R15.06-2012 standard is an adoption of pre-existing robot safety standards such as ISO 10218 which focuses on providing HRC cell designers with appropriate guidelines to help them improve the safety of collaborative cells. It provides some of the safety measures that should be used to ensure a safe collaborative environment, such as safety stop, hand guiding, speed and separation monitoring, power and force limiting. The main risk factors in the standards, ranked by importance, are severity, exposure, and avoidance. These factors provide a more detailed representation of risk compared to the flat values based on severity and probability. By taking into account these factors, a comprehensive assessment of risk level can be conducted, resulting in categorizations such as negligible, low, medium, high, and very high. Standards provide suggestions also on how using these factors during the design of an HRC cell to reduce the risks. Focusing on a robot control perspective, safety measures are about assigning tasks to the robot and controlling its motions in order to reduce the risk of collisions with the operator while keeping an low cycle time.

HRC-oriented works usually focus on general design issues (see, e.g., [15]) and on sub-problems such as scheduling human and robot actions (e.g., [3]), or cooperative planning at a symbolic level (e.g., [7,27]). Several works have investigated novel models and planning paradigms to enhance *human awareness* and realize more effective and safe robot behaviors in collaborative settings [6,11]. A central aspect is the capability of enriching a symbolic task planning level with geometric information suitable to reason about spatial qualities of planned motions. The work [22] although not strictly correlated with human-robot interaction, shows an interesting combination of task and motion planning suitable to improve task planning with geometric reasoning and plan feasible motion trajectories. The interesting aspect of this approach is the pursuit of a "shared perspective" allowing task planners to reason on more realistic (geometry-aware) models. This is especially relevant in HRC where an explicit reasoning on the *qualities* of robot motions is crucial to enable safer and more effective behaviors. The study [21] supports modeling multi-agent decision problems for human-aware task allocation. It uses a hierarchical task planning formalism to reason about structured decomposition and combined behaviors of multiple agents (e.g., human and robot) working together towards shared objectives (e.g., a production task [17]). However, this approach lacks explicit consideration of time constraints and action durations, limiting the effectiveness of generating optimal collaborative plans in terms of cycle time. The work [7] advances human-aware optimization by considering stochastic human dynamics. It creates a hierarchical model from single-agent demonstrations to capture task temporal relationships and prevent spatial conflicts in robot actions. Notably, this work combines safety and efficiency to improve planning decisions. The study [19] also combines stochastic human occupancy data in hierarchical collaborative modeling. It addresses task allocation, enabling robots to consider time constraints and human behavior for safe, efficient trajectory planning. The approach achieves a balance between safety and efficiency based on the agents' occupancy volumes. Another work [14] enhances simultaneous human-robot behaviors using a synergy matrix for parallel tasks. It employs a multi-objective planning formalism that successfully balances optimization and minimizing delays due to potential human-robot collisions. The study [16] introduces a new sequencing planning algorithm that optimizes cycle time while considering resource and safety constraints. Notably, it provides detailed safety reasoning by including constraints related to assembly/disassembly components. In manufacturing-oriented methods, [12] optimizes the ergonomics of the human worker by using an online workflow scheduler. [9,10] proposed a TAMP framework for planning and executing tasks using first-order logic graphs. A contingent-based approach was proposed in [1,2] to deal with uncertainty on the outcome of actions. Although the mentioned works effectively address human-robot interaction issues, none of them rely on an explicit model of risk supporting parametrized reasoning about safety. The majority of works dealing with safety mainly consider the expected spatial occupancy of human workers to make safer decisions about robot actions. Namely, such works mainly focus on deciding *which* tasks assign to the robot taking into account expected spatial requirements to reduce the expected cycle time. A finer-grained model of risk and collaboration dynamics could instead support more detailed decisions taking into account also *how* to execute robot tasks in order to mitigate potential risks.

The novel contribution of our work thus concerns the design of a detailed risk-aware model of collaborative tasks, based on existing safety standards, and the development of a multi-objective optimization approach to HRC. To the best of our knowledge, this is the first work investigating the use of a risk model extracted from safety standards within a multi-objective task planning framework. More specifically, task and motion planning rely on a timeline-based framework, called PLATINUm [24,25]. PLATINUm complies with the formal characterization proposed in [8] which takes into account *temporal uncertainty*, and has been successfully applied in real-world manufacturing scenarios [14,19].

3 A Risk Awareness Model for Task Sequencing

Drawing from existing standards and concepts, our risk awareness model aims to enhance safety and efficiency in HRC domains beyond typical design guidelines. Our proposal offers a versatile model applicable to various operational scenarios, with a particular emphasis on addressing task sequencing planning challenges. We propose a model consisting of 5 risk factors describing the safety characteristics of HRC cells to endow cobots with *sophisticated* risk awareness.

The *intrinsic risk* value reflects the severity concept and indicates the inherent risk level of a specific task. It is determined by properties intrinsic to the task itself and independent of its execution method. Factors considered in this value include characteristics of the tool (e.g., hot, powered, sharp, etc.), compliance and design of the robot (e.g., sharp or pointy parts), the use of protective equipment by the human operator (e.g., glasses, gloves), etc. The *geometry risk* depends on which trajectory the robot is following to carry out its task. The risk value of a trajectory is assigned based on how likely it is for that trajectory to intersect a human body part and which part it is likely to intersect. *Speed risk* is also a risk value, and it intuitively depends on how fast the robot is carrying out its task. *Collision risk* is a risk value that depends on how likely it is for the human and robot to be in each other's way. It mainly depends on the tasks being carried out by the human and the robot at the same time. This is a value that most closely translates to exposure. Impact probability is high not only if the human and the robot are close together but also when they accomplish a strictly collaborative task. The *human unpredictability* represents the concept of avoidance (along with geometry risk and movement speed). It is higher when the system is less certain about the human's movement or intention. As an example, human unpredictability can encapsulate human expertise. When an operator is highly skilled in HRC, then we assume uncertainty to be low.

The risk factors can be utilized to define a heuristic function with values like worst case, best case, or more advanced logic. Typically, these values are determined by knowledge and production engineers, analyzed within the [0,1] range, and carry equal weight in calculating the final risk value for a task. The final risk value is derived by multiplying all the values together. More formally, we consider a scenario with a set of tasks A that a robot might carry out. The i-th task in A can be executed with a set of distinct trajectories T_i (each of them with its own risk factor $t_{i,j}$). Each of the trajectories can be followed with one of the speed values $v_{i,j,k}$ in the set $V_{i,j}$. Each of the tasks in A has a set of tasks P_i that can be carried out concurrently by the human (each of those with its own risk factor $p_{i,n}$). Each of the tasks in A also has its own level of intrinsic risk a_i. The operator that is being considered in the given scenario can have one of many uncertainty values inside U. For simplicity, we assume here that the human uncertainty u depends on her expertise level. To quantify a comprehensive risk value, we consider the following cumulative risk formula: $R_{i,j,k,n} = a_i \cdot t_{i,j} \cdot v_{i,j,k} \cdot p_{i,n} \cdot u$. Each one of the values assigned to risk factors depends on its own specific environmental properties. A task planner can reduce the intrinsic risk by swapping the current task with another, or it can change the trajectory and speed to reduce geometry risk and movement speed respectively; it can reallocate the tasks to reduce collision probability; it can communicate with the operator to reduce the human unpredictability. Being aware of the risk is a crucial feature for both efficiency and safety. A robot that is aware of safety can make better-informed decisions on when to perform safety stops, thus reducing the number of unnecessary interruptions. Risk awareness at planning time may allow a scheduler to come up with a plan leading to fewer safety stops and slow-downs, greatly improving efficiency and safety. We will explore a proof of concept implementation of our planner later. This model uses parameters that are generic and can be potentially applied to any scenario since they are derived from the first principles considered in the standards. These parameters satisfy our requirements of flexibility and generality quite well and they can be easily reinterpreted in specific situations to better fit a given scenario.

4 Evaluation and Experiments

We developed a risk-aware task sequencing system for human-robot collaboration using PLATINUm. The system incorporates a risk model based on the 5 risk factors introduced above and employs the heuristics defined by these factors. To this aim, we implemented a *RiskAssessmentPlanner* with a customized search strategy, which plays a crucial role in estimating the potential of partial plans and determining the most promising options to achieve production goals. As in [14], we pursue a Pareto optimization approach to strike a balance between safety and efficiency. This approach avoids extreme solutions that completely eliminate collaboration between humans and robots in order to minimize risk.

In order to showcase the most relevant features of our planner as well as assess the effectiveness of our model and the associated heuristics, we consider an experimental scenario inspired by an EU-funded research project, called shareWork in which a human and a robot collaborate to build a mosaic [13]. The first relevant feature of our planner is its ability to sensibly assess the risk of a situation using our model. This would prove the effectiveness of our model and enable us to build on these foundations and come up with new experiments and implementations that could further expand the possibilities of HRC. Then, the experiments are to show that our planner is effectively leveraging its risk awareness to strike a good balance between safety and efficiency (i.e., reducing risks while improving productiveness). The experiments show also the ability of our planner to take good decisions and to come up with better plans when can decide with more freedom. This is particularly interesting for considering the impact of such freedom with respect to planning time and scalability. Finally, we aim at showing how task assignment change according to different human expertise levels. A smart collaborative robot should be aware indeed of who it is collaborating with and act accordingly [26].

The experiment considers a scenario based on a series of Pick-And-Place operations, one of the most common operations in HRC. A robot and a human collaborate to sort a set of cubes from a common working area into separated heaps, based on the cube's material. Three different kinds of cubes are considered: foam, wood, and hot metal cubes. Each of these materials has different intrinsic risks associated with them. A foam cube is safe for the robot to move because it is soft, so the possibility of accidentally hitting the human using a foam cube has a low severity. Wood is firm and it is riskier to move since hitting the human with a wooden cube might hurt them, especially if there is no personal protective equipment. Hot metal cubes are very risky for the robot to move because hitting the human with it would almost probably cause injuries.

	PickPlaceFoam			PickPlaceWood			PickPlaceMetal		
Intrinsic risk	0.2			0.5			0.8		
Geometry risk	Legs	Chest	Head	Legs	Chest	Head	Legs	Chest	Head
	0.2	0.4	0.7	0.2	0.4	0.7	0.2	0.4	0.7
Movement speed	Slow	Regular	Fast	Slow	Regular	Fast	Slow	Regular	Fast
	0.1	0.3	0.6	0.1	0.3	0.6	0.1	0.3	0.6
Impact probability	PPF	PPW	PPM	PPF	PPW	PPM	PPF	PPW	PPM
	0.6	0.4	0.4	0.4	0.6	0.4	0.4	0.4	0.6
Human uncertainty	Unexp.	Experienced		Unexp.	Experienced		Unexp.	Experienced	
	0.7	0.3		0.7	0.3		0.7	0.3	
Risk Value	The product of values chosen above			The product of values chosen above			The product of values chosen above		
Makespan	Evaluated using trajectory and speed data			Evaluated using trajectory and speed data			Evaluated using trajectory and speed data		

Fig. 1. Example of precompiled data

Since there are only Pick-And-Place operations, this means that the intrinsic risk of a given task only depends on what kind of cube is being moved. Each Pick-And-Place task can be carried out with one of three different trajectories and one of three different speeds. The shortest trajectory is one that moves the cube over the working area and it could collide with the operator's head. This trajectory is the fastest one but it is also the riskiest one. There is also a trajectory that could collide with the human's chest, so it is safer but also slower. The last trajectory moves the cube under the working area, where it is slower and safer as it could only collide with the operator's feet. Each of these trajectories can be followed at a medium or high speed, thus making it faster but also riskier. In the experiment, we consider one human with one of two levels of experience that are the same for the whole duration of the plan. The expertise level can be either high or low. All these settings can be combined in order to allow the planner to face harder decisions in which it has to choose between prioritizing either safety or efficiency. We define a sensible set of precompiled data that would be fed into the algorithm. A good set of values is crucial in getting the best results out of a planner. Figure 1 shows an example of precompiled data used with the planner and to reflect the risk estimation that we implicitly do as humans.

In order to assess our search strategy (*RiskAssessmentSearchStrategy*), we compared its performance with other planners considering different heuristics: a strategy giving the highest priority to safety (*RiskSearchStrategy*), i.e., assigning task to minimize safety and not considering makespan; a strategy giving highest priority to makespan (*MakespanStrategy*), i.e., assigning the task to minimize makespan and not considering safety; a classical depth-first strategy (*PlannerSearchStrategy*). In order to have an accurate depiction of the quality of our tools, we decided to make many variations of the experiment with varying properties and see how the final result changed. We consider 6 cubes of each kind for each experiment and different variations. Namely, we consider a varying amount of shared tasks between humans and robots, i.e., variations where 0 tasks are shared and both the human and the robot have been allocated 3 cubes of each kind. For each of the variations with different amounts of shared cubes, there are 2 cases: one where the human is experienced and one where the human is inexperienced. This allows us to see how the planners react to every combination of parameters. There are 2 expertise values and 4 shared amounts, so there are 8 plan variations. Each of the 4 tested strategies will run 5 times per variation, for a total of 160 runs. We performed the experiment multiple times, with different sets of precompiled data. The tests were performed on a workstation with a Ryzen 3600 CPU and 16 GB RAM.

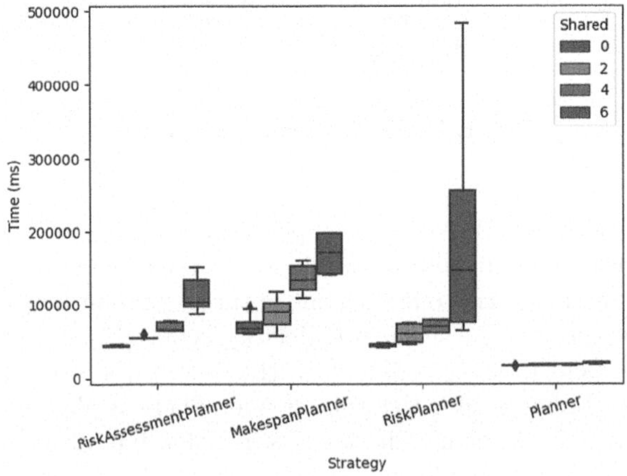

Fig. 2. Planning time against strategy and shared tasks

The analysis of the collected results provided several general indications. Our model and heuristic function enable our planner to make more informed decisions in task allocation. Our planner consistently keeps the risk level low and, whenever possible, significantly enhances efficiency. On the one hand, our planner performs a task assignment that adapts based on the assessed risk and, when feasible, it manages to avoid collisions even in the presence of task allocation constraints. On the other hand, the planner provides the best trade-off between safety and efficiency as it is able to reduce the risk of collisions, leading to the highest possible efficiency. Figure 2 shows how the computation time spent to find a suitable plan changes for each strategy, and how increasing the number of shared tasks between the human and the robot has an impact on planning time. Simple Planner (DFS) is the fastest one as it is only concerned with plan suitability and is not aware of safety or efficiency. Other strategies have similar behavior with RiskPlanner being the one suffering the most from a higher amount of shared cubes. Indeed, a higher shared amount of boxes entails PLATINUm to work with a much higher branching factor, i.e., taking more time to find a solution.

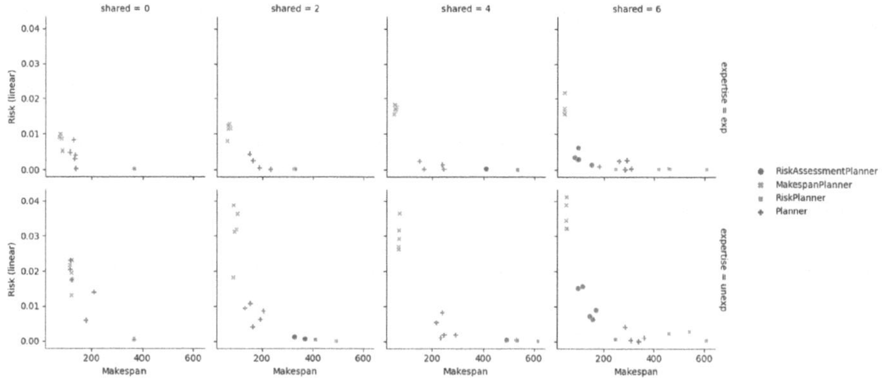

Fig. 3. Scatterplot with risk and makespan, grouped by expertise and shared tasks

In Fig. 3, the scatter plot illustrates how each strategy balances risk and efficiency. Data points are categorized by expertise and shared values. The significant aspect is the right side, indicating RiskAssessmentPlanner's consistent superiority in safety over MakespanPlanner, while being faster than both Planner and RiskPlanner. For less experienced humans, RiskAssessmentPlanner acknowledges heightened risk, producing safer yet slower plans in response.

5 Discussion and Conclusions

Our new risk model allows diverse perspectives in planning systems. Interestingly, in certain experiments, the basic DFS Planner matches the RiskAssessmentPlanner in safety, e.g., with 6 tasks for experienced workers, and is even safer (though less efficient) for inexperienced workers. This outcome is not entirely unexpected. Although our approach enhances flexibility, optimizing efficiency and safety without compromise remains a challenge, aiming for a trade-off between the two. Indeed, RiskAssessmentPlanner slightly sacrifices safety for better efficiency while DFS Planner tends to show lower collaboration levels (averaging 5.4 tasks out of 6 assigned to the robot). This discrepancy is because DFS Planner's risk is less impacted by an inexperienced operator who contributes very little (averaging 20 time units compared to the robot's 80). Moreover, despite entailing a higher risk, RiskAssessmentPlanner performs better in terms of collision prevention and task assignment. Therefore, safety levels might appear comparable in plans but there are substantial differences in both average and peak risks. RiskAssessmentPlanner excels in collision handling and task allocation. DFS Planner maintains low risk by mostly assigning tasks to the robot (i.e. limiting collaboration). RiskAssessmentPlanner prioritizes effective task allocation and collision management, even if it means sacrificing efficiency, especially in constrained scenarios. Integrating diverse risk factors into task sequencing models enables intelligent risk-aware control in human-robot collaboration (HRC). This leverages a state-of-the-art planning framework for risk-aware task sequencing. Our approach showcased efficacy and adaptability across varied simulations, achieving a favorable productivity-safety equilibrium. In future steps, we plan to validate our method in real collaborative setups and incorporate real-time user feedback. This feedback could account for factors like operator perceived risk.

Acknowledgments. Authors are partially supported by Italian MUR within the Fit4MedRob PNRR project (PNC0000007).

References

1. Akbari, A., Diab, M., Rosell, J.: Contingent task and motion planning under uncertainty for human-robot interactions. Appl. Sci. **10**(5), 1665 (2020)
2. Akbari, A., Muhayyuddin, Rosell, J.: Knowledge-oriented task and motion planning for multiple mobile robots. J. Exp. Theor. Artif. Intell. **31**(1), 137–162 (2019)
3. Casalino, A., Zanchettin, A.M., Piroddi, L., Rocco, P.: Optimal scheduling of human-robot collaborative assembly operations with time petri nets. IEEE Trans. Autom. Sci. Eng. **18**, 70–84 (2019)
4. Cesta, A., Orlandini, A., Bernardi, G., Umbrico, A.: Towards a planning-based framework for symbiotic human-robot collaboration. In: 2016 IEEE 21st International Conference on Emerging Technologies and Factory Automation (ETFA) (2016)
5. Cesta, A., Orlandini, A., Umbrico, A.: Fostering robust human-robot collaboration through AI task planning. Procedia CIRP **72**, 1045–1050 (2018). 51st CIRP Conference on Manufacturing Systems
6. Chen, J., Ding, Y., Xin, B., Yang, Q., Fang, H.: A unifying framework for human-agent collaborative systems-part I: element and relation analysis. IEEE Trans. Cybern. **52**, 138–151 (2022)
7. Cheng, Y., Sun, L., Tomizuka, M.: Human-aware robot task planning based on a hierarchical task model. IEEE Robot. Autom. Lett. **6**(2), 1136–1143 (2021)

8. Cialdea Mayer, M., Orlandini, A., Umbrico, A.: Planning and execution with flexible timelines: a formal account. Acta Informatica **53**(6–8), 649–680 (2016)
9. Darvish, K., Bruno, B., Simetti, E., Mastrogiovanni, F., Casalino, G.: Interleaved online task planning, simulation, task allocation and motion control for flexible human-robot cooperation. In: 2018 27th IEEE International Symposium on Robot and Human Interactive Communication (RO-MAN) (2018)
10. Darvish, K., Simetti, E., Mastrogiovanni, F., Casalino, G.: A hierarchical architecture for human-robot cooperation processes. IEEE Trans. Robot. **37**(2), 567–586 (2021)
11. Ding, Y., Xin, B., Chen, J., Yang, Q., Fang, H.: A unifying framework for human-agent collaborative systems-part II: design procedure and application. IEEE Trans. Cybern. **52**(11), 11990–12002 (2022)
12. Evangelou, G., Dimitropoulos, N., Michalos, G., Makris, S.: An approach for task and action planning in human-robot collaborative cells using AI. Procedia CIRP **97**, 476–481 (2021)
13. Faroni, M., et al.: A layered control approach to human-aware task and motion planning for human-robot collaboration. In: 29th IEEE International Conference on Robot and Human Interactive Communication, pp. 1204–1210 (2020)
14. Faroni, M., Umbrico, A., Beschi, M., Orlandini, A., Cesta, A., Pedrocchi, N.: Optimal task and motion planning and execution for multiagent systems in dynamic environments. IEEE Trans. Cybern. (2023)
15. Huck, T.P., Münch, N., Hornung, L., Ledermann, C., Wurll, C.: Risk assessment tools for industrial human-robot collaboration: novel approaches and practical needs. Saf. Sci. **141**, 105288 (2021)
16. Lee, M.L., Behdad, S., Liang, X., Zheng, M.: Task allocation and planning for product disassembly with human-robot collaboration. Robot. Comput. Integr. Manuf. **76**, 102306 (2022)
17. Lemaignan, S., Warnier, M., Sisbot, E.A., Clodic, A., Alami, R.: Artificial cognition for social human-robot interaction: an implementation. Artif. Intell. **247**, 45–69 (2017)
18. Ogenyi, U.E., Liu, J., Yang, C., Ju, Z., Liu, H.: Physical human-robot collaboration: robotic systems, learning methods, collaborative strategies, sensors, and actuators. IEEE Trans. Cybern. **51**, 1888–1901 (2019)
19. Pellegrinelli, S., Orlandini, A., Pedrocchi, N., Umbrico, A., Tolio, T.: Motion planning and scheduling for human and industrial-robot collaboration. CIRP Ann. **66**, 1–4 (2017)
20. Plaku, E., Kavraki, L.E., Vardi, M.Y.: Motion planning with dynamics by a synergistic combination of layers of planning. IEEE Trans. Robot. **26**(3), 469–482 (2010)
21. de Silva, L., Lallement, R., Alami, R.: The HATP hierarchical planner: formalisation and an initial study of its usability and practicality. In: 2015 IEEE/RSJ International Conference on Intelligent Robots and Systems (IROS), pp. 6465–6472 (2015)
22. Srivastava, S., Fang, E., Riano, L., Chitnis, R., Russell, S., Abbeel, P.: Combined task and motion planning through an extensible planner-independent interface layer. In: 2014 IEEE International Conference on Robotics and Automation (ICRA), pp. 639–646 (2014)
23. Tsarouchi, P., Makris, S., Chryssolouris, G.: Human-robot interaction review and challenges on task planning and programming. Int. J. Comput. Integr. M. **29**(8), 916–931 (2016)

24. Umbrico, A., Cesta, A., Cialdea Mayer, M., Orlandini, A.: Integrating resource management and timeline-based planning. In: The 28th International Conference on Automated Planning and Scheduling (ICAPS) (2018)
25. Umbrico, A., Cesta, A., Cialdea Mayer, M., Orlandini, A.: PLATINUm: a new framework for planning and acting. In: Esposito, F., Basili, R., Ferilli, S., Lisi, F. (eds.) Advances in Artificial Intelligence (AI×IA 2027). Springer International Publishing (2017). https://doi.org/10.1007/978-3-319-70169-1_37
26. Umbrico, A., et al.: Design of advanced human-robot collaborative cells for personalized human-robot collaborations. Appl. Sci. **12**(14), 6839 (2022)
27. Wang, W., Li, R., Diekel, Z.M., Jia, Y.: Robot action planning by online optimization in human-robot collaborative tasks. Int. J. Intell. Robot. App. **2**(2), 161–179 (2018)

Artificial Intelligence at Manufacturing Process Level

Use of Artificial Intelligence at the Level of Manufacturing Processes

Panagiotis Stavropoulos(✉), Alexios Papacharalampopoulos, and Dionysios Christopoulos

Laboratory for Manufacturing Systems and Automation (LMS), Mechanical Engineering and Aeronautics Department, University of Patras, 26504 Patras, Greece
pstavr@lms.mech.upatras.gr

Abstract. Artificial Intelligence (AI) is an ensemble of technologies that have diverse capabilities. As a matter of fact, it appears that most of them are quite diverse and useful, also when applied in the field of manufacturing processes. However, the interactions between the agents involved in relevant manufacturing applications often is not straightforward, as the environment can be quite harsh and dynamic. Herein, a mapping is sketched between AI territories and manufacturing process-related operations. To this end, different aspects of the processes are considered, such as monitoring, modelling, optimization, design and preparation, as well as control, twinning and operation. For each one of these cases, AI technologies and workflows are matched, leading to a roadmap for future research. In this path, there are various specific challenges, many of which are mentioned here, as well. Specific examples from both conventional and non-conventional processes are considered, as well as focused illustrations of the possibilities that AI technologies can offer. In addition, the different manufacturing attributes, as guided by different strategies, are discussed in the context of the aforementioned concept. These include, among others, sustainability, human centricity and quality management. Finally, the role of services is discussed under the current context.

Keywords: Artificial Intelligence · Manufacturing Processes · modelling · monitoring · optimization

1 Introduction

AI is crucial for improving production systems, enabling the integration of new applications, supporting complex decision-making [1]. It is essential for strengthening manufacturing processes by streamlining operations and enhancing overall efficiency through several techniques.

Machine learning (ML), subset of AI, has transitioned from research to practical use, leveraging larger datasets for knowledge extraction, while demonstrating effectiveness in numerous tasks like object detection and image recognition, to name a few [2]. Fuzzy logic (FL) incorporates multiple values in human decision-making, utilizing fuzzy inference and fuzzy rules [3]. Expert systems (ESs), utilize the knowledge of human specialists to solve problems [4] and Heuristic algorithms (H-ALG) imitate the human's

A. Wagner et al. (Eds.): ESAIM 2023, LNME, pp. 157–166, 2024.
https://doi.org/10.1007/978-3-031-57496-2_16

implicit analysis [5]. All the aforementioned techniques, individually or in combination, enhance manufacturing domain in many aspects of the manufacturing processes, like monitoring, modelling, optimization, design and preparation, control, twinning and operation in general. Hereafter, a mapping is drawn between operations connected to manufacturing processes and AI regions, drafting a roadmap for the future.

The established terms are grouped in a specific way, since AI according to books has its fluctuations in books (Fig. 1). It is worth noting that "AI" is a term found even in 1800's books, however, the classification is as heuristic per use (i.e. [6]).

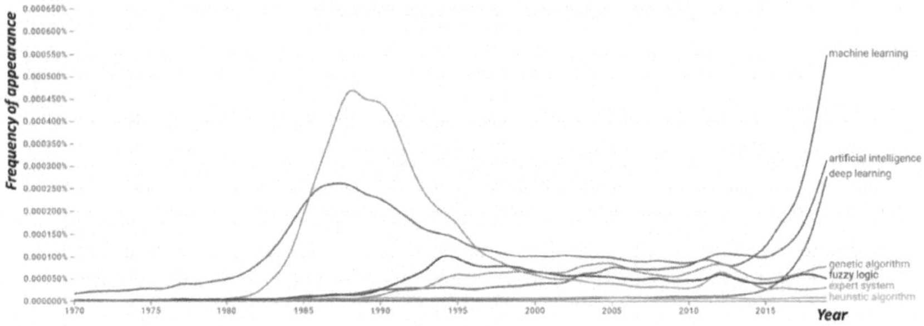

Fig. 1. Frequency of appearance of AI-related terms in books, according to Google N-Gram [7].

2 State of the Art

2.1 Techniques

A rough quantitative analysis in Scopus reveals the distribution of works pertaining to AI and manufacturing processes. Figure 2 summarizes the number of works found. This analysis is prone to a lot of criticism, as every single abstract has to be checked. Hereafter, a qualitative analysis can be found. These four categories have been chosen due to their frequency of occurrence in literature [8, 9] their historical and operational significance [10] and the state-of practice, mainly in EU research projects [11].

X\Y	optimization	monitoring	control	modelling
machine learning	295	301	454	345
expert systems	57	35	178	82
fuzzy logic	70	39	167	95
heuristic algorithms	80	2	38	15

Fig. 2. Heatmap based on the number of Scopus findings after the query X & Y & "manufacturing process" on July 7th, 2023.

Expert Systems (ES). A rule-based ES is implemented to address problems with defect prevention and monitoring in green sand molding casting process. A fuzzy ES based on particle-swarm optimization (PSO) is used to forecast the mechanical characteristics of injection-molded components. To find the best process parameters, optimization using a PSO technique is also carried out, with the goal to optimize the tensile strength of the molded product [12]. An ES, linking with libraries for features, materials, abrasives, and machine settings and using a feature-based approach to gather design specifications, is used for Abrasive WaterJet Machining (AWJM) optimization. The ES provides optimization of machining cycle time, cost, and cutting rate by modifying machine parameters, reducing machining time and cost by 10% compared to results from a CNC abrasive waterjet [13]. A data-driven smart agent system is presented, which identifies optimal Additive Manufacturing (AM) methods for digital designs on a computer network. The system consists of a dynamic machine identification system, an Internet of Things (IoT) simulator, and an ES based on Artificial Neural Networks (ANN), recommending suitable section planes for additive repair using Laser powder bed fusion (LPBF) is introduced.The volume cut-off, necessary support structures, and the effect on holes or functional regions are all taken into consideration while evaluating these planes [14]. A dual polarity phase shift soft switch to enhance the dynamic properties of the power supply's voltage and current waveforms, is utilized in a control system. A specialized welding ES, using the Newton interpolation algorithm, is developed to automatically adjust welding process parameters and provide features such as parameter unification and automatic generation. The effectiveness of the cold-welding power supply, combined with the ES, is verified [15]. To forecast surface roughness in green AWJM, a thorough ES is described that combines a Takagi-Sugeno-Kang (TSK) fuzzy model with subtractive clustering (SC). The integrated expert system is intended to improve the surface roughness prediction's accuracy and dependability in the AWJM process for green composites [16].

Machine Learning (ML). ML lays the groundwork for ongoing quality monitoring and management of highly sophisticated and dimensional operations [17]. In powder-bed fusion additive manufacturing (PBFAM), acoustic emission (AE) signals are interpreted using ML techniques like spectral convolutional neural networks (SCNN) [18]. In continuous casting of steel, a multilevel NN (MNN) is employed for temperature pattern analysis using backpropagation NN (BPNN) and fuzzy NN for decision-making [17]. Statistical models like response surface methodology (RSM), ANN, and Support Vector Machines (SVM) were utilized to analyze the impact of cutting factors on surface roughness and workpiece vibration during stainless steel boring [19]. SVM, AdaBoost, and Random forests can be utilized to predict the surface roughness of printed parts under various process conditions [20]. An ANN model was utilized to establish a relationship between printing parameters and the development of the extrudate in 3D printing. The model was trained using data from extrudate-based experiments, and its predicted extrudate geometry closely matched the actual geometry [21]. Digital twin-based forward and inverse models, employing recurrent neural networks (RNNs) such as long short-term memory (LSTM), bidirectional LSTM (BiLSTM), and gated recurrent units (GRUs), ensure precise prediction of key performance indicators (KPIs) based on process parameters (PPs) and vice versa in Laser PBFAM [22].

Fuzzy Logic (FL). L. A. Zadeh developed fuzzy logic (FL) for the first time in 1965. Additionally, it was widely employed in inference systems subsequently. These systems are frequently used to optimize production system and process characteristics [23]. The

MATLAB fuzzy logic Toolbox is used to create an implementation of the Mamdani inference system (IS), a kind of fuzzy IS. This implementation's goal is to assess the risks connected to scope, scheduling, cost, and quality elements in research and development projects related to AM [24]. For optimizing the EDM process for milling SS304, a FL IS were utilized to produce a Multi-Response Performance Index (MRPI) (ASTM A240) [23]. FL inference is applied in a control approach to enhance fabrication accuracy in Wire Arc AM (WAAM). This technique adjusts deposition parameters based on geometrical flaws in previous layers, establishing interlayer closed-loop control (ICLC). High height accuracy is achieved in the produced cuboid blocks the suggested ICLC technique was found to improve deposition continuity in WAAM, facilitating process automation in robotic manufacturing [25]. Prediction of flexural strength of Acrylonitrile butadiene styrene (ABS) parts using Fused Deposition Modeling (FDM) is attempted, proposing a FL model. The model considered factors such as layer thickness, raster width, raster angle, and orientation angle. The results demonstrated that the FL models provided accurate predictions of flexural strength with a maximum 3.62% error [26]. A FL model was used with the Mamdani technique to forecast the quality laser cutting (LC), based on cutting parameters such laser power, cutting speed etc. in CO2 LC of alloy AA6061-T6 aluminum process. The FL model accurately predicted the top and bottom kerf width, as well as the kerf deviation in CO2 LC, as confirmed by the experimental results [27]. An adaptive neuro-fuzzy inference system (ANFIS) is proposed to establish a flexible model linking input process data and parameters with the shape of weld beads in Gas Metal Arc Welding (GMAW). The Adaptive neuro fuzzy inference system (ANFIS) accurately represents weld bead geometry using GMAW parameters like voltage, welding speed, and wire feed rate [3].

Heuristic Algorithms (H-ALG). ANN and support vector machine (SVM) models were compared for quality prediction for milling operations. Optimization algorithms, including Genetic Algorithm (GA), Particle Swarm Optimization (PSO), and simulated annealing (SA), were tested to determine optimal cutting parameters, indicating that combining an ANN model with subsequent PSO yielded the best results [28]. An automated Smart Manufacturing (SM) heuristic-based scheduling architecture blending Blockchain and Digital Twin (DT) Technologies for providing legal identity verification and protect device privacy is proposed. The cloud layer manages data scheduling for SM and deploys DT [29]. The limitations of Selective Laser Melting (SLM) in relation to upright-printed scaffold shapes is investigated proposing two distinct design heuristics for slanting and cluttering. The heuristics are evaluated through multiple iterations [30]. An adaptive large neighborhood search with a memetic algorithm (MA) incorporating a GA is proposed for reducing travel distance and workpiece temperature during laser cutting [5]. Electrical Discharge Machining (EDM) process is simulated and optimized using ANN and PSO algorithm, determining the optimal inputs for optimizing Material Removal Rate, Tool Wear Rate and Surface Roughness. The PSO algorithm achieved effective optimization with less than 4% error. The proposed BPNN modeling accurately replicated the EDM process with less than 1% error [31]. A GA for reducing support volume by beam removal and optimizing inner and exterior support structures is used in AM, achieving less material consumption, and outperforming other existing strategies in material consumption [32]. To improve the speed tracking capability of a wire-feeder system in GMAW, a fuzzy fractional controller was used through PSO and numerical simulations. The wire-feeder system's reference tracking, torque disturbance rejection, and resilience against model uncertainties were improved [33]. Figure 3 hereafter summarizes the workflows in literature. It is worth noting that additional algorithms can be considered AI, such as control and (meta)modelling techniques.

2.2 Manufacturing Attributes

The use of AI in an industrial setting can aid in learning more and identifying the necessary conditions for achieving Zero Defect Manufacturing (ZDM). The research into an AI application for ZDM shows that AI not only makes ZDM easier but also complements conventional quality techniques and helps build more robust manufacturing processes [34]. Sustainable manufacturing calls for particular procedures and tackling related issues through all-encompassing strategies. An organized and practical holistic approach, centered on ZDM, is recommended to support sustainable manufacturing. The use of technologies like AI, ML, DTs, and knowledge-based tools is also highlighted. Both at the management level and at the production level, these technologies offer precise event prediction, that enables excellent decision-making at the appropriate time, enhancing efficiency and sustainability, some highly important topics, as manufacturing systems nowadays use a lot of wastes of any kind, including materials, energy, other natural resources, and manufacturing time, which is contrary to the sustainability aims [35].

3 Discussion

3.1 Current Challenges and Future Roadmap

The extraction of the challenges is not a straightforward procedure. A very careful review of the literature mentioned above, as well as the investigation of additional works, in terms of business (i.e. [36]), implementation-wise (i.e. [37]) and from the perspective of data (e.g. [38]) is required.

So, to begin with, it can be claimed that getting the right data is a common challenge when employing ML in manufacturing. This restriction results from elements that have a substantial effect on how well ML algorithms function, such as the availability, quality, and content of industrial data. High-dimensional data that can add duplicate and irrelevant information and impair the performance of learning algorithms are potential challenges associated to the dataset. Furthermore, most ML methods now only work with continuous and nominal data types [39]. The development of manufacturing technologies, including cyber-physical systems, the IIoT, and AI, has accelerated the transition of production infrastructures into interconnected networks that include automation equipment, services, and businesses.

As a result, this growth creates a difficulty by increasing the need for interoperability at different levels within the manufacturing environment [38]. Another challenge is addressed since two important aspects must be considered to determine if DT-driven supervised ML is feasible to use in industrial practice. These considerations include the

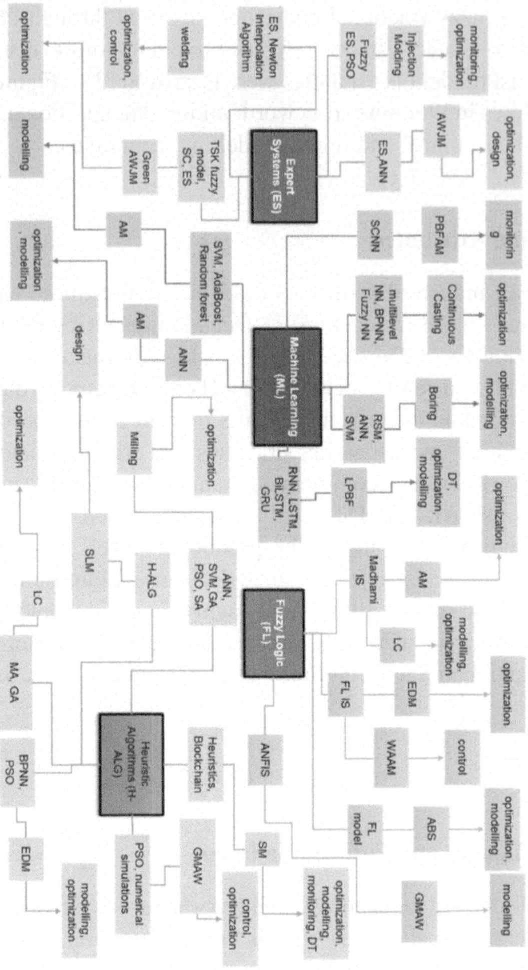

Fig. 3. Matching procedures with AI techniques [Technique—Methods —Process→Operation].

costs involved in creating and incorporating new features into the DT system, as well as the precision of the DT model and simulation services [1]. Additional challenges are pertaining to metamodelling functions; i.e. automated estimation of ARX model order [40].

There are a number of approaches that may be taken to deal with the difficulties of using AI in the manufacturing sector. These include enhancing data management practices to guarantee data availability and quality, using feature selection and dimensionality reduction techniques to handle high-dimensional data, researching techniques for handling data types other than continuous and nominal, implementing interoperability standards to enable seamless system communication, assessing the costs and precision of AI systems, and continuously enhancing their accuracy. For the manufacturing sector to successfully address these issues and utilize AI, collaboration between domain experts and practitioners is essential. In addition, the co-existence of different techniques seems to be mandatory; especially towards the implementation of digital twins (Fig. 4). Such is the case, after all, with current frameworks [41].

3.2 The Role of Services

The concept of services is also pertaining to AI-related challenges, with a two-way dependence [42, 43]. In addition to this, manufacturing companies strive for customer loyalty, profitability, and sustainability by offering personalized products with added services. The corresponding approach, known as Product-Service System (PSS), creates value through collaboration and addresses economic, environmental, and social concerns. The goal is to provide a bundled package that includes product performance and usage, rather than just the product itself [44]. The integration of PSS and AI holds great potential for enhancing manufacturing processes. Incorporating AI, manufacturers could utilize customer data, through ML algorithms that collect and analyze purchase history and browsing behavior or Natural Language Processing (NLP) techniques that can be employed to analyze customer feedback, reviews, and social media data, so as to optimize the implementation of PSS, resulting in customized and tailored product configurations, and meeting customers' needs and preferences. AI's predictive analytics capabilities could allow for the anticipation of maintenance requirements, the identification of potential product failures and the automation of decision-making. The combination of AI and PSS may help at satisfying the principles of circular economy as well. With the development of sustainable and circular manufacturing practices, products and services are designed for longevity and reusability, achieving reduced waste, improved environmental impact, and enhanced social responsibility.

3.3 Addressing Strategies

A Digital Twin based on various types of AI can be highly useful. Decision making, predictions, automation and control can all be supported by a single piece of software. In addition, optimization with respect to concepts such as quality management (defect recognition and prevention) as well as sustainability (interpreted as the reduction in use of resources and emissions) can all be supported. The overall implementation cost, however, is still unknown, since it is depending on the data and the equipment. To this end, harmonization with human intervention can be an additional target. This knowledge exchange, together with desired visualization (and manipulation [45]) of information, will lead to addressing the latest strategies, such as Industry 5.0.

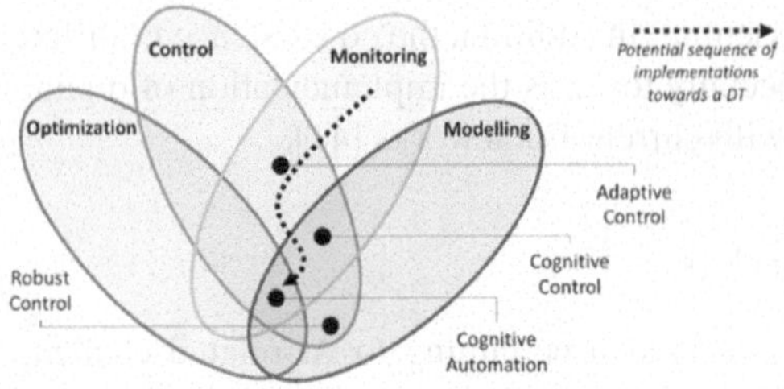

Fig. 4. Operations and demand for AI techniques towards digital twins (DT's).

Acknowledgements. The current work has been partially funded by EU project AIRISE, under contract 101092312.

References

1. Alexopoulos, K., Nikolakis, N., Chryssolouris, G.: Digital twin-driven supervised machine learning for the development of artificial intelligence applications in manufacturing. Int. J. Comput. Integr. Mnauf. **33**(5), 429–439 (2020)
2. Fahle, S., Prinz, C., Kuhlenkötter, B.: Systematic review on machine learning methods for manufacturing processes - identifying artificial intelligence methods for field application. In: 53rd CIRP Conference on Manufacturing Systems, pp. 413–418. Elsevier B.V., Bochum (2020)
3. Foorginejad, A., Azargoman, M., Mollayi, N., Taheri, M.: Modeling of weld bead geometry using adaptive neuro-fuzzy inference system (ANFIS) in additive manufacturing. Appl. Comput. Mech. **6**(1), 160–170 (2020)
4. Zhao, P.: Intelligent injection molding on sensing, optimization, and control. In: Advances in Polymer Technology, pp. 1–22 (2020)
5. Bologna, F., Tannous, M., Romano, D., Stefanini, C.: Automatic welding imperfections detection in a smart factory via 2-D laser scanner. Manuf. Processes **73**, 948–960 (2022)
6. AIMNET white paper. https://www.aim-net.eu/. Accessed 11 July 2023
7. Google N-Gram site. https://books.google.com/. Accessed 11 July 2023
8. Rai, R., Tiwari, M. K., Ivanov, D., Dolgui, A.: Machine learning in manufacturing and industry 4.0 applications. Int. J. Prod. Res. **59**(16), 4773–4778 (2021)
9. Fahle, S., Prinz, C., Kuhlenkötter, B.: Systematic review on machine learning (ML) methods for manufacturing processes–Identifying artificial intelligence (AI) methods for field application. Procedia CIRP **93**, 413–418 (2020)
10. Chryssolouris, G.: Manufacturing Systems: Theory and Practice. Springer, New York (2006). https://doi.org/10.1007/978-1-4757-2213-0
11. EFFRA Site. https://www.effra.eu/artificial-intelligence-and-manufacturing-share-your-experience-guidance. Accessed 18 Aug 2023
12. Kramar, D., Cica, D.: Predictive model and optimization of processing parameters for plastic injection moulding. Materiali in Tehnologije **51**(4), 597–602 (2017)
13. Amalnik, M.S.: Design and manufacturing optimization of abrasive water jet machining using expert system. ADMT **12**(1), 101–114 (2019)
14. Ganter, N.V., Hoppe, L.V., Dünte, J., Gembarski, P.C., Lachmayer, R.: Knowledge-based assistance system for part preparation in additive repair by laser powder bed fusion. In: Proceedings of the Design Society, pp. 1381–1390. Cambridge University Press (2022)

15. Huang, D.S., Huang, Z.K., Hussain, A.: Intelligent computing methodologies. In: ICIC Proceedings, Part III, vol. 10956. Springer, Cham (2018). https://doi.org/10.1007/978-3-031-13832-4
16. Jagadish, Bhowmik, S., Ray, A.: Prediction of surface roughness quality of green abrasive water jet machining: a soft computing approach. Intell. Manuf. **30**(8), 2965–2979 (2019)
17. Cemernek, D.: Machine learning in continuous casting of steel: a state-of-the-art survey. Intell. Manuf. **33**(6), 1561–1579 (2022)
18. Shevchik, S.A., Masinelli, G., Kenel, C., Leinenbach, C., Wasmer, K.: Deep learning for in situ and real-time quality monitoring in additive manufacturing using acoustic emission. IEEE Trans. Ind. Inf. **15**(9), 5194–5203 (2019)
19. Venkata Rao, K., Murthy, P.B.G.S.N.: Modeling and optimization of tool vibration and surface roughness in boring of steel using RSM, ANN and SVM. Intell. Manuf. **29**(7), 1533–1543 (2018)
20. Li, Z., Zhang, Z., Shi, Z., Wu, D.: Prediction of surface roughness in extrusion-based additive manufacturing with machine learning. Robot. Comput.-Integr. Manuf. **57**, 488–495 (2019)
21. Lao, W., Li, M., Wong, T.N., Tan, M.Z., Tjahjowidodo, T.: Improving surface finish quality in extrusion-based 3D concrete printing using machine learning-based extrudate geometry control. Virtual Phys. Prototyping **15**(2), 178–193 (2020)
22. Papacharalampopoulos, A., Michail, C.K., Stavropoulos, P.: Manufacturing resilience and agility through processes digital twin: design and testing applied in the LPBF case. In: 9th CIRP Global Web Conference, pp. 164–169. Elsevier B.V., Rio Patras (2021)
23. Ubaid, A.M., Dweiri, F.T., Aghdeab, S.H., Al-Juboori, L.A.: Optimization of electro discharge machining process parameters with fuzzy logic for stainless steel 304 (ASTM A240). Manuf. Sci. Eng. Trans. ASME **140**(1) (2018)
24. Moreno-Cabezali, B.M., Fernandez-Crehuet, J.M.: Application of a fuzzy-logic based model for risk assessment in additive manufacturing R&D projects. Comput. Ind. Eng. **145**, 106529 (2020)
25. Li, Y., Li, X., Zhang, G., Horváth, I., Han, Q.: Interlayer closed-loop control of forming geometries for wire and arc additive manufacturing based on fuzzy-logic inference. Manuf. Process. **63**, 35–47 (2021)
26. Mamo, H.B., Tura, A.D., Johnson Santhosh, A., Ashok, N., Kamalakara Rao, D.: Modeling and analysis of flexural strength with fuzzy logic technique for a fused deposition modeling ABS components. Mater. Today Proc. **57**, 768–774 (2022)
27. Parthiban, A., Sathish, S., Ravikumar, R., Prakash, P.: Experimental investigation and fuzzy logic modelling of CO2 laser cutting parameter for AA6061-T6 sheet. Int. J. Rapid Manuf. **9**(1), 48–57 (2020)
28. Coppel, R., Abellan-Nebot, J.V., Siller, H.R., Rodriguez, C.A., Guedea, F.: Adaptive control optimization in micro-milling of hardened steels—evaluation of optimization approaches. Int. J. Adv. Manuf. Technol. **84**, 2219–2238 (2016)
29. Chen, H., Jeremiah, S.R., Lee, C., Park, J.H.: A digital twin-based heuristic multi-cooperation scheduling framework for smart manufacturing in IIoT environment. Appl. Sci. **13**(3), 1440 (2023)
30. Kostadinov, A., Yan, L., Teo, A.Q.A., O'Neill, G.: Slanted and cluttered: solving deficiencies in SLM-manufactured lattice geometries. Mater. Des. **211**, 110130 (2021)
31. Moghaddam, M.A., Kolahan, F.: Modeling and optimization of the electrical discharge machining process based on a combined artificial neural network and particle swarm optimization algorithm. Scientia Iranica **27**(3), 1206–1217 (2020)
32. Vaissier, B., Pernot, J.P., Chougrani, L., Véron, P.: Genetic-algorithm based framework for lattice support structure optimization in additive manufacturing. Comput. Aided Des. **110**, 11–23 (2019)

33. Hamouda, N., Babes, B., Kahla, S., Hamouda, C., Boutaghane, A.: Particle swarm optimization of fuzzy fractional PDμ+I controller of a PMDC motor for reliable operation of wire-feeder units of GMAW welding machine. Przeglad Elektrot. **96**(12), 40–46 (2020)
34. Leberruyer, N., Bruch, J., Ahlskog, M., Afshar, S.: Toward Zero defect manufacturing with the support of artificial intelligence—insights from an industrial application. Comput. Ind. **147**, 103877 (2023)
35. Psarommatis, F., Bravos, G.: A holistic approach for achieving sustainable manufacturing using Zero Defect Manufacturing: a conceptual Framework. In: 55th CIRP Conference on Manufacturing Systems, pp. 107–112. Elsevier B.V., Oslo (2022)
36. Burström, T., Parida, V., Lahti, T., Wincent, J.: AI-enabled business-model innovation and transformation in industrial ecosystems: a framework, model and outline for further research. J. Bus. Res. **127**, 85–95 (2021)
37. Bokrantz, J., Subramaniyan, M., Skoogh, A.: Realising the promises of artificial intelligence in manufacturing by enhancing CRISP-DM. Prod. Planning Control 1–21 (2023)
38. Zeid, A., Sundaram, S., Moghaddam, M., Kamarthi, S., Marion, T.: Interoperability in smart manufacturing: research challenges. Machines **7**(2), 21 (2019)
39. Wuest, T., Weimer, D., Irgens, C., Thoben, K.D.: Machine learning in manufacturing: advantages, challenges, and applications. Prod. Manuf. Res. **4**(1), 23–45 (2016)
40. Papacharalampopoulos, A.: Investigating data-driven systems as digital twins: numerical behavior of Ho–Kalman method for order estimation. Processes **8**(4), 431 (2020)
41. Stavropoulos, P., Papacharalampopoulos, A., Stavridis, J., Sampatakakis, K.: A three-stage quality diagnosis platform for laser-based manufacturing processes. Int. J. Adv. Manuf. Technol. **110**, 2991–3003 (2020)
42. Nicoletti, B., Appolloni, A.: Artificial Intelligence for the management of Servitization 5.0. Sustainability **15**(14), 11113 (2023)
43. Manser Payne, E.H., Dahl, A.J., Peltier, J.: Digital servitization value co-creation framework for AI services: a research agenda for digital transformation in financial service ecosystems. J. Res. Interact. Mark. **15**(2), 200–222 (2021)
44. Zheng, P., Lin, T.G., Chen, C.H., Xu, X.: A systematic design approach for service innovation of smart product-service systems. J. Clean. Prod. **201**, 657–667 (2018)
45. Maass, W., Storey, V.C., Lukyanenko, R.: From mental models to machine learning models via conceptual models. In: Augusto, A., Gill, A., Nurcan, S., Reinhartz-Berger, I., Schmidt, R., Zdravkovic, J. (eds.) BPMDS/EMMSAD -2021. LNBIP, vol. 421, pp. 293–300. Springer, Cham (2021). https://doi.org/10.1007/978-3-030-79186-5_19

Metal Forming Process Efficiency Improvement Based on AI Services

Fernando Boto[1], Daniel Cabello[2], Juan Antonio Ortega[1], Blanca Puigjaner[2], and Asier Alonso[1(✉)]

[1] TECNALIA, Basque Research and Technology Alliance (BRTA), Mikeletegi Pasealekua 7, 20009 Donostia-San Sebastián, Spain
{fernando.boto,jantonio.ortega,asier.alonso}@tecnalia.com

[2] DENN, Carrer Pintor Vila Cinca, 30, 08213 Polinyà, Barcelona, Spain
{dcabello,bpuigjaner}@denn.es

Abstract. In this work, the application of Artificial Intelligence (AI) and Machine Learning (ML) techniques in the field of metal forming processes in Shear Forming and Spinning machines are explored. The main objective is to improve the quality of the parts produced and the efficiency of these processes through the implementation of predictive models and online value-added services.

Firstly, different methods for the analysis and evaluation of the quality of manufactured parts are presented. Additionally, predictive models for online failure detection are developed, based on historical and real-time data, which helps prevent failures and reduce production costs.

Furthermore, the challenge of detecting changes in the input material, which can have a significant impact on process outcomes, is addressed.

Lastly, the implementation of an algorithm towards "zero defects" is proposed to achieve optimal conditions in the metal forming process.

The described approaches enable customers of the incremental forming machine manufacturer to access a diverse range of services associated with the implemented methods. ...

Keywords: Metal forming · Machine Learning · Process data · Industry 4.0

1 Introduction

Market demands are increasingly emphasizing the importance of easily reconfigurable machines capable of manufacturing a variety of different products. In this context servitization in manufacturing has become more relevant as Artificial Intelligence (AI) has advanced and demonstrated its ability to enhance production processes. By combining AI with manufacturing equipment, companies can achieve greater efficiency, optimize production, reduce downtime, and improve product quality. This trend is becoming more feasible due to the digitization of machine tools to improve production efficiency and product quality [1]. AI

A. Wagner et al. (Eds.): ESAIM 2023, LNME, pp. 167–176, 2024.
https://doi.org/10.1007/978-3-031-57496-2_17

techniques do not mandate an extensive comprehension of the underlying physical mechanisms that link the variables of the target process to its responses [2]. Notably, supervised and unsupervised machine learning algorithms also provide a standardized procedure for constructing models that ensure quality [3].

In recent years, incremental forming processes [4–6] have gained attention. These processes involve the massive deformation of a flat or tubular workpiece in three dimensions, but with the application of localized forces, in contrast to processes such as deep drawing, where an entire sheet deforms at once. Symmetric incremental forming processes involve rotation around an axis, along which the material is incrementally deformed at specific points.

One of the most strategically important processes for the sector is spinning. The metal sheet is progressively deformed against the mold, either using a manual tool or through small rollers. In this process, the thickness of the sheet remains constant, while the diameter decreases. A variant of the process is called shear spinning. In this process, the diameter of the component remains the same as that of the starting sheet, but the thickness of the sheet is reduced according to the following geometric expression: $t = t_o c.sin\alpha$.

Some papers suggest that data analytics and machine learning have several benefits in incremental metal forming. For example in [7] discusses the potential for AI to improve the single point incremental forming (SPIF) process. [8] specifically looks at the use of machine learning algorithms to predict forming force. Other studies [9], achieve optimal process parameters to minimize surface roughness, which they validate through experimental validation. These works reveal the interest in servitizing this technology in the process itself, but there are few works that show an evidence of using them as a service in real application. For example [10] and [11] identify opportunities for future research. These papers demonstrate the potential of AI servitization in incremental metal forming to optimize processes and improve product quality. In the study cited as [12], the authors address the deficiency in achieving effective closed-loop control over product properties in metal forming processes. This necessitates rigorous experimentation and dedicated effort in developing the corresponding process models.

This paper proposes different and separated services based on data analytics and machine learning as part of a holistic approach adopted by the manufacturer. This work aims for minimal configuration and training of the models to facilitate the seamless transfer of the models to the machine. The strategies implemented in this paper are integrated into a cloud-based web application and an edge application to manage data from an entire fleet of rotary machines. The solution serves as a suite of added services for the manufacturer's customers, providing them with control over the process. The generated services encompass data-driven machine learning tools for quality assessment and online control.

In this study, the proposed approach intertwines smoothly with the Digital Twin framework [13], as it combines the synergy of online quality assessment and control services. By integrating these services, a dynamic and accurate representation of the manufacturing process can be maintained virtually, allowing for real-time monitoring, optimization, and informed decision-making. The authors' exploration of these services not only aligns with the broader concept of a Digital Twin but also underscores its pivotal role in fostering efficient and responsive production environments.

This work is based on the most relevant variables for extracting knowledge from the mentioned processes. Figure 1 shows the signal acquired for each single part. The process variables used are the following: head rotation speed, work carriage speed (X, Z), work carriage force (X, Z), work carriage position (X, Z) and momentary motor torque.

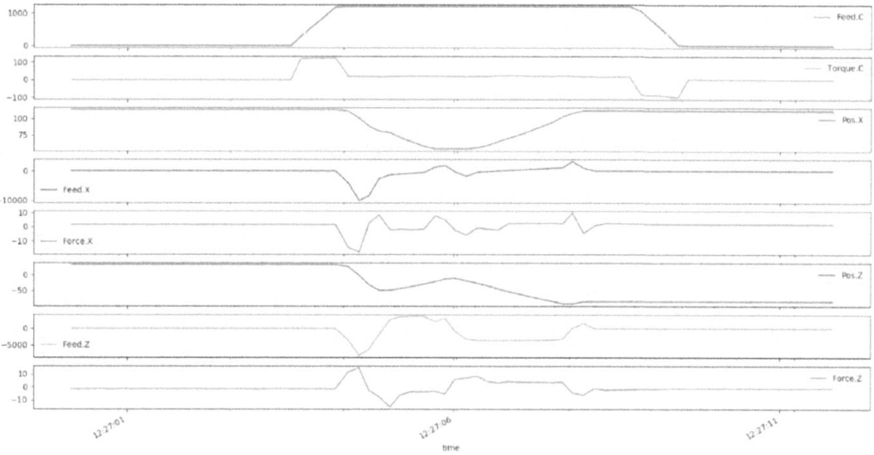

Fig. 1. Incremental forming signals for a single part. Head rotation speed (Feed.C), carriage speed (Feed.X, Feed.Z), carriage force (Force.X, Force.Z), carriage position (Pos.X, Pos.z) and torque (Torque.C)

It should be noted that the process is always very stable, and therefore, the force, position, and velocity signals uniquely characterize a reference. However, tooling dilations can introduce instabilities in the process. Additionally, material changes or accidental/deliberate adjustments to the preform can result in errors and machine damage. Quality assurance or fracture detection, as well as preform changes, are crucial for the customers of the incremental forming machine manufacturing company. Hence, this study presents results in these areas to provide customers with added value to the data extracted from the machines.

The paper is structured as follows. First, a section focuses on the detailed description of the implemented methods and services. Section 3 presents the results and discussion pertaining to the implemented services, followed by the presentation of conclusions.

2 Smart Services for Metalforming

2.1 Quality Assessment of Produced Parts

An algorithm has been developed through the analysis of velocity, position, and force data. The strategy consists of three steps: data selection for training, generation of a normality model based on a supervised approach, and finally, classification of the parts using statistical analysis (Fig. 2).

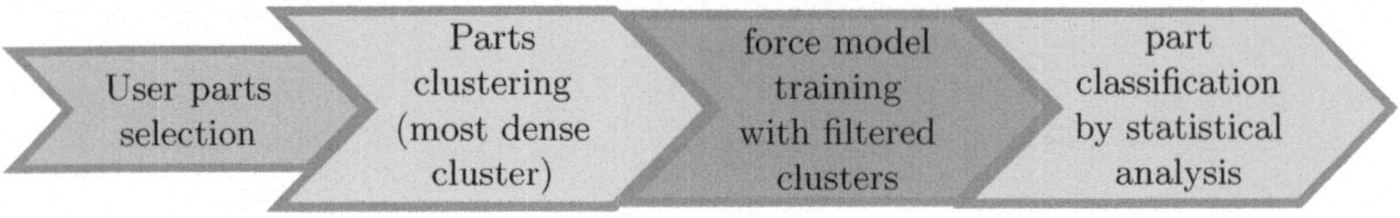

Fig. 2. Quality assessment algorithm flow diagram

In the data selection process, the user input is taken into account. Therefore, initially, the expert selects the initial parts with a known correct behavior. Once this selection is made, to ensure the purity of the dataset, an automatic clustering is performed to discard isolated parts in low-density clusters. Affinity Propagation, a versatile clustering technique, was employed [14], allowing for cluster formation without excessive parameterization and eliminating the need to predefine the number of clusters.

From the aforementioned process variables (position, velocity, and force), various descriptors have been extracted that identify the process cycle for each part. These descriptors are related to the cycle time, idle points during execution, and roller advancement. Taking into account these descriptors the clustering is performed and most dense cluster is selected. Each cluster represents a set of parts that exhibit similar behavior. The assumption behind this filtering is that the parts with favorable outcomes form the densest cluster. The parts within this cluster are the ones selected to train the normality model.

The strategy taken in this work involves developing a normality model by utilizing a regressor that establishes correlations between different process parameters. Instead of solely relying on the observation of individual values, this approach analyzes the relationships and dependencies among the parameters. By constructing the regressor, the interaction between the parameters is captured, establishing a reference for normal behavior. This approach allows for the identification of significant deviations in parameter values and detection of anomalous patterns that may indicate undesirable operation or production system failures. The regression paradigm utilized in this case is the Random Forest bagging method [15]. A hyperparameterization of 100 estimators is employed, without a maximum depth for the estimator, and utilizing a quadratic loss. For generating the model in each execution, the service trains a distinct model where each sample represents a relationship between position, velocity, and force at each timestamp of the part's fabrication. To generate the model, all timestamp are used as samples from the selected pieces in the previous step, with the same specified hyperparameters. As a result, we obtain a model that represents the strong correlation between the inputs (position and velocity) and the output (force) of the process. During the model training process, the Mean Absolute Error (MAE) is measured, which serves as the metric for establishing decision boundaries for classifying well or poorly manufactured parts.

The generated normality pattern is used to statistically quantify the percentage of timestamp instances in a cycle or manufactured part that deviate from the pattern. For this purpose, the widely used empirical rule in statistics has been employed. The empirical rule, also known as the 68-95-99.7 rule [16], is a fundamental property of the normal distribution, which is one of the most important statistical distributions in probability theory. This rule states that in a normal distribution, approximately 68% of the data falls within one standard deviation of the mean (σ).

The training error, as said before, derived from the normality pattern is utilized to construct a Gaussian distribution and establish quality levels based on the empirical rule (σ). This approach incorporates the estimation error of a new piece by incorporating the mean absolute error into the distribution. As a result, it becomes possible to analyze the quality of the new piece within the established framework.

In the context of generating an error distribution and establishing thresholds, the MAE provides insight into the dispersion and magnitude of errors in the model predictions. If thresholds are set based on the MAE, they can identify outliers or be used to assess the quality of predictions. Therefore, in this implementation, instead of using σ, the MAE has been employed as a boundary, and to align with the empirical rule, three boundaries are established: MAE, $2 * MAE$, and $3 * MAE$.

2.2 Optimal Control for Quality Improvement

The proposal presented in this work aims to modify a crucial parameter of the shear forming process to maintain the stability. The mold of the machine undergoes thermal deformations that require a geometric adjustment. The hypothesis followed to improve adaptation to this phenomenon is the revision of force behavior. To achieve this, the normality pattern strategy described earlier has been employed, which relates position and velocity to force. As mentioned earlier, the normality pattern establishes the relationship between force, velocity, and position as $force(velocity, position)$ based on well-produced pieces. However, when deformation is produced incrementally, there is an error in estimating the force compared to the actual force. The mean absolute error of the executed piece serves as an indicator that can be utilized as a parameter for correcting the process.

Taking a reference piece as a basis, the model generates a pattern. When a new piece with different conditions is introduced, a larger error is observed in the model, corresponding to the required geometric adaptation in the process.

This error is transformed into an offset that corrects the machine mold position, ensuring that no defects occur in the resulting piece. Without this correction, the piece would otherwise be rejected due to poor quality. Based on the generated force error, a mold position correction will be applied to maintain the initial force pattern.

2.3 Real Time Control

This section presents two methods implemented for both the spinning and shear forming processes, where machine problems are detected in real time. The common point in these new approaches is the generation of indicators that allow detecting these problems by establishing an appropriate threshold in each case.

First we present an algorithm for the automatic detection of changes in the input preformed sheet during incremental deformation processes. The changes that have been taken into account are: material changes (aluminum, steel, etc.), thickness changes, and material loading errors (no sheet loaded). Our approach is based on the analysis of the mean of the force signal of multiple executions for each preformed sheet change. By calculating the mean of these executions, we obtain a representative model for each type of material. Then, when classifying a new piece, we calculate the coefficient of determination to assess its similarity to the previously generated models. A predefined threshold allows us to determine which material change has occurred, offering precise and reliable detection. The threshold is determined empirically to achieve an overall satisfactory classification error. The algorithm exhibits a high level of versatility, ensuring accurate performance across various machine and process types, including both spinning and shear forming processes.

Second an algorithm for real-time prediction of part failure in an incremental deformation process is described. The algorithm utilizes an indicator based on Principal Component Analysis (PCA) applied to five process variables related to vibrations, force, and torque.

PCA is applied to the collected data using a specific time window until time T. Two principal components are extracted, capturing the most significant variations in the dataset. These components serve as the basis for constructing the failure indicator. The resulting indicator is computed as:

$$\sqrt{{C_1}^2 + {C_2}^2} \text{ where } C_1 \text{ and } C_2 \text{ are the first two components of the reduction} \tag{1}$$

If this indicator exceeds a predetermined threshold, a failure is detected.

To establish an optimal threshold for failure detection, an automated algorithm is developed. A labeled dataset comprising various part references and their corresponding failure statuses is used for training. The algorithm iteratively searches for the threshold that maximizes the classification performance based on F_1 score, achieving a balance between false positives and false negatives.

3 Results and Discussion

Firstly, the results of the quality assessment algorithm utilizing the three-step approach are presented. The filtered executions with the clustering method are selectively chosen to generate the normality model, while the remaining executions are discarded.

Secondly, we are analyzing the performance and benefits of the developed normality pattern. The selected strategy to establish the relationship between position, velocity, and force is random forest regression [15]. As the configuration of the normality pattern varies in each case, we can observe different levels of reliability in the regression model. However, it is worth noting that the regressor demonstrates a strong correlation among the variables, achieving errors ranging between 3% and 5%.

The final quality classification is derived through the Gaussian analysis described earlier. This approach enables us to effectively measure the severity of any instances of poor quality that may arise, providing users with relevant information through graphical results. This allows for a comprehensive assessment of the detected issues and facilitates informed decision-making.

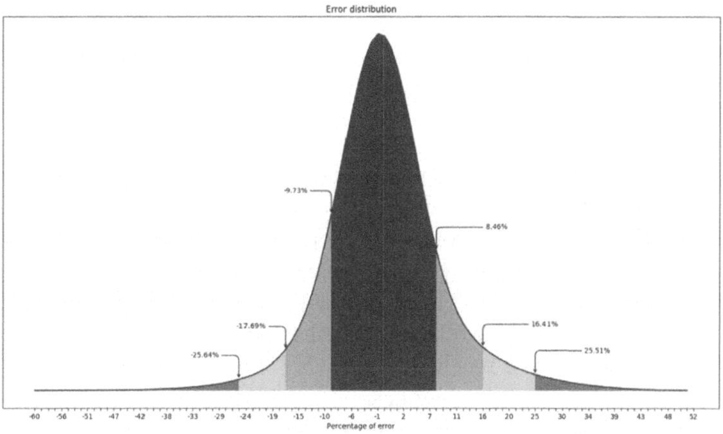

Fig. 3. Statistical and visual analysis of quality parts

Figure 3 illustrates an example of the bell curve generated during the training of the normality pattern. This bell curve enables the assessment of defect severity in a given piece. By comparing the error of a new piece to the pattern, its deviation is visually represented on the Gaussian bell curve. An initial analysis performed with this service has demonstrated a classification accuracy of 93% for correctly classified pieces.

Regarding the process optimization, we have conducted a preliminary analysis that enables us to provide a control value based on the force applied to the material. This preliminary analysis allows us to establish initial guidelines for optimizing the process, providing a valuable reference for monitoring and adjusting the process parameters in real-time.

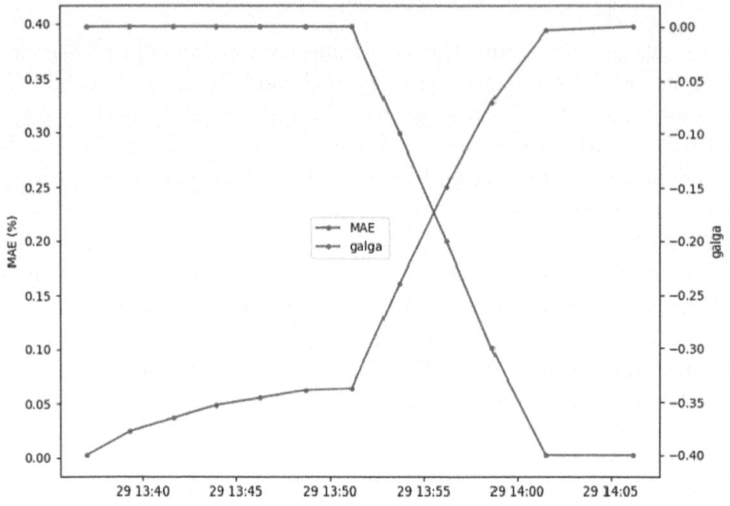

Fig. 4. Shear forming optimal process control with a force modeling approach

When thermal expansion occurs in the mold, a geometric compensation becomes necessary. Typically, the user operating the machine performs this compensation by adjusting the position of the mold along the Z-axis. This adjustment is often done through a trial-and-error process or guided by the user's experience. The underlying hypothesis of the implemented approach is that the force exerted by the tool on the material correlates with the required compensation. In order to validate this approach, a test was conducted where the position of the mold was incrementally modified to compensate for thermal distortion. In the Fig. 4, it can be observed that this hypothesis is validated, as the error generated by the normality pattern model is inversely proportional to the required geometric compensation to be applied to the machine.

When it comes to real-time control services, emphasizing their low-latency requirements is crucial. Therefore, the two approaches proposed here have been implemented as services within an edge application. This integration enables the control system to effectively utilize real-time algorithmic information, facilitating timely adjustments on the machine. This enables rapid decision-making and ensures efficient machine control in real-time scenarios.

To evaluate the preform detection effectiveness of our algorithm, we conducted comprehensive tests using different types of materials, including aluminum, iron, and others. The results showed high accuracy in detecting material changes, with minimal rates of false positives and false negatives. The accomplishment is a method that enables the machine to halt its operations prematurely, thus preventing any damage to its tools and avoiding unnecessary financial expenses.

The classification accuracy is above 97%, indicating a high level of reliability for integration into the actual process. The change in the preform is detected before 15% of the execution time of the part, avoiding reaching the end of the execution with a loading error.

This result is achieved by implementing an automatic threshold, as described in Sect. 2.3. In the edge application, the user provides feedback to gradually enhance the algorithm's performance.

This method offers significant potential for enhancing process monitoring and reducing production losses.

Regarding the real-time rupture detection, the results demonstrate that the incremental PCA-based indicator, combined with the optimized threshold determination algorithm, yields a high classification accuracy exceeding 95% in the spinning process. This indicates a significant level of reliability and potential for integration into real-time production processes. The proposed method enables timely detection of part failure, allowing for prompt intervention and prevention of economic losses.

4 Conclusions

The objective of this work is to offer a comprehensive suite of AI services for metal forming processes.

It is worth highlighting that all the described methods are implemented within a service-oriented platform. This platform allows customers of the incremental forming machine manufacturer to access a wide range of services related to the implemented methods. This remarkable achievement is the result of years of collaboration between the machine manufacturer and a research center, both sharing a unified vision of translating AI methods into practical industry applications. This serves as an excellent example of successful knowledge transfer from applied research to the industrial sector.

In future work, it is important to reflect on the type of data used and explore opportunities for enhanced supervision. The incorporation of active learning techniques could prove beneficial in this regard, as ongoing research is currently investigating its potential.

References

1. Liu, C., Zheng, P., Xu, X.: Digitalisation and servitisation of machine tools in the era of Industry 4.0: a review. Int. J. Prod. Res. **61**(12), 4069–4101 (2021). https://doi.org/10.1080/00207543.2021.1969462
2. Stavropoulos, P., Sabatakakis, K., Papacharalampopoulos, A., Mourtzis, D.: Infrared (IR) quality assessment of robotized resistance spot welding based on machine learning. Int. J. Adv. Manuf. Technol. **119**, 1785–1806 (2022)

3. Stavropoulos, P., Papacharalampopoulos, A., Sabatakakis, K.: Online quality inspection approach for submerged arc welding (SAW) by utilizing IR-RGB multi-modal monitoring and deep learning. In: Kim, KY., Monplaisir, L., Rickli, J. (eds.) International Conference on Flexible Automation and Intelligent Manufacturing. LNME, pp. 160–169. Springer, Cham (2022). https://doi.org/10.1007/978-3-031-18326-3_16
4. Li, Y., et al.: A review on the recent development of incremental sheet-forming process. Int. J. Adv. Manuf. Technol. **92**(5–8), 2439–2462 (2017). https://doi.org/10.1007/S00170-017-0251-Z
5. Ajay, C.V., Boopathi, C., Kavin, P.: Incremental sheet metal forming (ISMF): a literature review. In: AIP Conference Proceedings, vol. 2128, p. 030012. AIP Publishing LLC (2019)
6. Trzepieciński, T.: Recent developments and trends in sheet metal forming. Metals **10**(6), 779 (2020)
7. Kumar, S.P., Elangovan, S., Mohanraj, R., Boopathi, S.: Real-time applications and novel manufacturing strategies of incremental forming: an industrial perspective. Mater. Today Proc. **46**, 8153–8164 (2021). https://doi.org/10.1016/J.MATPR.2021.03.109
8. Ostasevicius, V., Paleviciute, I., Paulauskaite-Taraseviciene, A., Jurenas, V., Eidukynas, D., Kizauskiene, L.: Comparative analysis of machine learning methods for predicting robotized incremental metal sheet forming force. Sensors **22**(1), 18 (2021). https://doi.org/10.3390/s22010018
9. Kurra, S., Rahman, N.H., Regalla, S.P., Gupta, A.K.: Modeling and optimization of surface roughness in single point incremental forming process. J. Market. Res. **4**(3), 304–313 (2015)
10. Harfoush, A., Haapala, K.R., Tabei, A.: Application of artificial intelligence in incremental sheet metal forming: a review. Procedia Manuf. **53**, 606–617 (2021). https://doi.org/10.1016/j.promfg.2021.06.061
11. Hartmann, C., Opritescu, D., Volk, W.: An artificial neural network approach for tool path generation in incremental sheet metal free-forming. J. Intell. Manuf. **30**(2), 757–770 (2016). https://doi.org/10.1007/s10845-016-1279-x
12. Polyblank, J.A., Allwood, J.M., Duncan, S.R.: Closed-loop control of product properties in metal forming: a review and prospectus. J. Mater. Process. Technol. **214**(11), 2333–2348 (2014)
13. Fuller, A., Fan, Z., Day, C., Barlow, C.: Digital twin: enabling technologies, challenges and open research. IEEE Access **8**, 108952–108971 (2020)
14. Frey, B.J., Dueck, D.: Clustering by passing messages between data points. Science **315**(5814), 972–976 (2007)
15. Goel, E., Abhilasha, E.: Random forest: a review (2017)
16. Huber, F.: A Logical Introduction to Probability and Induction. Oxford University Press, Oxford (2018)

Fabric Defect Detection and Localization

Filipe Oliveira, Davide Carneiro(✉), Hugo Ferreira, and Miguel Guimarães

INESC TEC, R. Dr. Roberto Frias, 4200 Porto, Portugal
davide.r.carneiro@inesctec.pt

Abstract. Quality inspection is crucial in the textile industry as it ensures that the final products meet the required standards. It helps detect and address defects, such as fabric flaws and stitching irregularities, enhancing customer satisfaction, and optimizing production efficiency by identifying areas of improvement, reducing waste, and minimizing rework. In the competitive textile market, it is vital for maintaining customer loyalty, brand reputation, and sustained success. Nonetheless, and despite the importance of quality inspection, it is becoming increasingly harder to hire and train people for such tedious and repetitive tasks. In this context, there is an increased interest in automated quality control techniques that can be used in the industrial domain. In this paper we describe a computer vision model for localizing and classifying different types of defects in textiles. The model developed achieved an mAP@0.5 of 0.96 on the validation dataset. While this model was trained with a publicly available dataset, we will soon use the same architecture with images collected from Jacquard looms in the context of a funded research project. This paper thus represents an initial validation of the model for the purposes of fabric defect detection.

Keywords: Fabric inspection · defect detection · object detection

1 Introduction

Quality inspection is a crucial aspect of the textile industry, ensuring that final products meet set quality, durability, and aesthetic benchmarks [1]. This process identifies and corrects defects or deviations in manufacturing, boosting customer satisfaction. Through meticulous inspections, manufacturers can pinpoint and resolve issues like fabric flaws, inconsistent dyeing, stitching problems, and sizing variations. It also aids in adhering to regulations, safety norms, and industry certifications. Furthermore, it enhances production efficiency by recognizing areas for improvement, reducing waste, and curbing rework. In a textile market driven by demanding standards, robust quality inspections are essential for securing customer loyalty, bolstering brand reputation, and achieving lasting success.

In the last years, Machine learning has been used in its different forms to address this challenge [2, 3]. This paper describes the first steps in developing an automated quality inspection system for the textile industry. The goal is to develop a model that can match or surpass human performance in defect detection in textiles, and that is able to not only

A. Wagner et al. (Eds.): ESAIM 2023, LNME, pp. 177–184, 2024.
https://doi.org/10.1007/978-3-031-57496-2_18

detect a defect but also classify and locate it in the fabric. This is paramount to, later in the pipeline, implement prescriptive analyses that can relate production factors with the prevalence of defects, their location and size, just to name a few.

While many of the existing approaches treat the problem as an image classification one, we treat it as an object detection one [4]. In the former, the output is only a class for the whole image. In the latter the output is a class and bounding box, i.e., the defect location on the image. This allows us to extract additional features from the detection, such as the defect size, shape and location, which are of utmost importance for more in-depth analyses of the causes of the defects.

This paper describes the methodology used to train a model for detecting defects in fabric, the dataset used and the results obtained. The model shows a maximum mAP@0.5 of 0.96 on the validation data at epoch 96, and an mAP@0.5 of 0.93 on the test data. While this is still an initial contribution, it shows that the architecture used is suitable for this kind of problem, and will be explored in the future as we collect our own images from production Jacquard looms.

2 Methodology

2.1 Data Preparation

To evaluate the proposed architecture and given that the image acquisition process is still underway, we use the publicly available MVTEC Anomaly Detection Dataset[1] [5]. This dataset offers a collection of more than 5000 high-resolution images, which are categorized into fifteen distinct object and texture categories with a primary focus on industrial inspection applications. Within each category, there are both defect-free training images and a test set that contains images showcasing a variety of defects, as well as images without any defects present. Additionally, the dataset includes pixel-precise annotations that identify and label all anomalies present in the images. Specifically, we used the *carpet* category which contains fabric with regular patterns that are similar to our intended application domain and have five different classes of defects: color, cut, hole, metal contamination and thread. Examples of fabric with defects and their corresponding masks are shown in Fig. 1. As can be seen, some of these defects are quite challenging to detect even by a trained human, due to their size and shape.

Given that this dataset is designed for benchmarking anomaly detection methods, some data preparation tasks were necessary. Specifically, pixel-level masks (Fig. 2) were converted into text annotations describing the bounding boxes of the defects, suitable to be used with the selected architecture. This was done, for each mask, by finding the minimum and maximum coordinates of the white pixels, both in the X and Y axes. Then the center coordinates of the bounding box were calculated, as well as its width and height. All these values are normalized to the dimensions of the image. Each image mask was thus converted to an annotation that contains the following: class label, x center, y center, width and height.

Given the relatively small size of the dataset, image augmentation was also performed including horizontal and vertical flips as well as rotations at 90°, 180° and 270°. Table 1 shows the original size of the dataset, per class, and the final size after data augmentation.

[1] https://www.mvtec.com/company/research/datasets/mvtec-ad/.

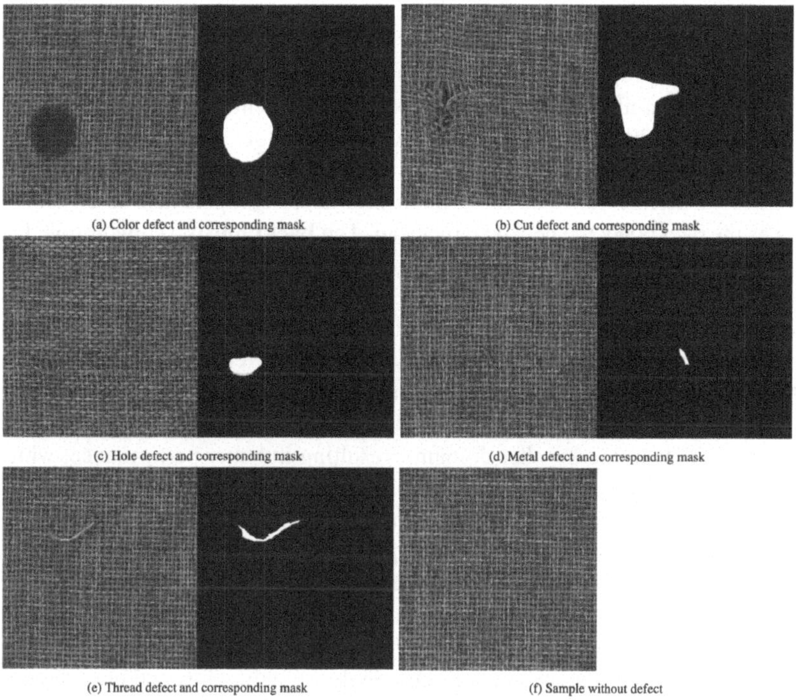

Fig. 1. Examples of images in the MVTEC dataset (carpet category) with different defects (color, cut, hole, metal, thread) and corresponding masks, and an image without defect.

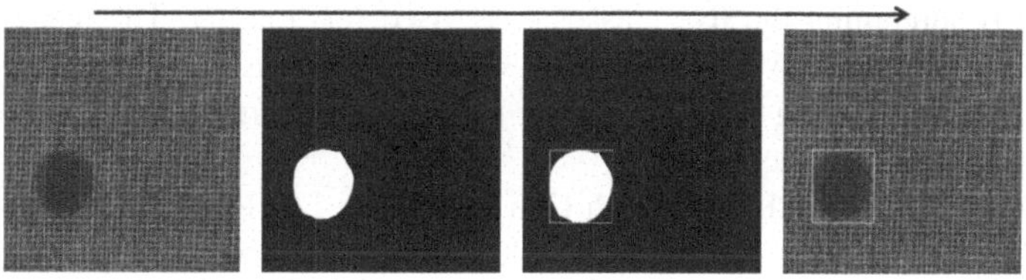

Fig. 2. Depicts the process of extracting the bounding boxes from the pixel-level masks.

Table 1. Size of the dataset, per class, before and after data augmentation.

Class	Original Size	After augmentation
Colour	19	152
Cut	17	136
Hole	17	136
Metal contamination	17	136
Thread	19	152
Total	*89*	*712*

2.2 Model Training and Evaluation

To implement the proposed defect detection system a YOLO-based architecture was used. YOLO is a family of architectures [6] for object identification based on an initial version released in 2016 [7]. YOLOv7, the version used in this work, can even perform real-time object detection [8].

YOLOv7's architecture has three main high-level components, described by the authors as the backbone, the neck and the head. The **backbone** is a convolutional neural network that aggregates picture pixels to generate characteristics at various levels of detail. The main framework is generally pre-trained on a classification dataset, such as ImageNet[2]. In the backbone of YOLOv7, which was based on YOLOv3, the Darknet53 backbone was substituted with the Resnet50-vd-dcn ConvNet. Resnet is a widely favored architecture, with more frameworks fine-tuned for its implementation, and it possesses fewer parameters compared to Darknet53. Using a better pretrained model has shown to improve transfer learning to other domains, resulting in better performance, which led to an enhancement in the mAP (Mean Average Precision).

The **neck** consists of a FPN (feature pyramid network), which combines and mixes the ConvNet layer representations, and then passes them to the prediction head. Another improvement in YOLOv7 was to move regularization from the backbone to the neck. Moreover, DropBlock regularization is used, which removes entire random blocks of features rather than individual ones, thereby improving generalization [9].

Finally, the **head** is responsible for defining the bounding box and providing a class prediction for each object found in the image. Additional information on the architecture of the model can be found in [8].

To ensure an unbiased evaluation of the model's performance on unseen data, the hold-out method was used. Given the relatively small size of the dataset, 90% of the images (640) containing the five classes were used for training. The rest of the data was used for validation (5%, 36 images), which allows assessing the generalization ability of the model during training, and for testing (5%, 36 images), to assess the final generalization ability of the model.

The model was trained on an 11th Gen Intel Core i7-11370H @ 3.30 GHz, with 16 GB RAM. It was configured to run for 100 epochs, with a batch size of 4. The size of the images was 640 × 640. The values for momentum and decay used were, respectively, 0.937 and 0.0005. The training of the model concluded after 3.8 h.

The evaluation of the model was based on the mAP, a commonly used performance metric in the field of object detection and information retrieval [10] used in evaluating architectures such as Faster R-CNN, MobileNet SSD, and YOLO. To compute it, we follow the standard methodology using in the Pascal Visual Object Classes (VOC) challenge [10]. The mAP measures the accuracy of a model by evaluating the precision and recall of its predictions, which are calculated through the Intersection over Union (IoU) metric for a given threshold (Eq. 1).

$$IoU(A, B) = \frac{A \cap B}{A \cup B} \tag{1}$$

[2] https://www.image-net.org/.

In this equation, A represents the set of pixels of the bounding box of the object while B represents the bounding box predicted by the model. If the IoU for a given prediction is above the threshold, it is marked as a True Positive. Otherwise, it is marked as a False Positive. The mAP is calculated at a given threshold, usually 0.5, in which case it is represented as mAP@0.5. The higher the threshold, the more robust the model is.

In this work, we additionally consider the mAP@[.5:.95], which represents the average mAP over different thresholds ranging from 0.5 to 0.95, using increments of 0.05. To compute mAP@[.5:.95], the precision-recall curve is first obtained by varying the threshold for predicting positive instances inside the provided range. Then, the area under the precision-recall curve is calculated, representing the average precision. Finally, the mAP is obtained by averaging the AP values across multiple classes or instances. The main advantage of mAP is that it provides a single numerical value to assess the overall performance of an object detection system.

3 Results

This section details the results of the evaluation of the trained model. Table 2 provides a summary of the statistics. The main indicator is the mAP@[.5:.95], which resulted in a value of 0.59. In what concerns the mAP@0.5, its highest value was obtained in the 96th epoch: 0.96. The corresponding values of Precision and Recall are, respectively, 0.94 and 0.96 and the mAP@0.95 was 0.93.

Table 2 also presents, on its right side, the results concerning losses, both for the training and validation sets. The Box-Loss measures how close the predicted bounding boxes are to the ground truth object. The Object-Loss measures the degree of confidence on the prediction. Finally, the Class-Loss measures the loss calculated on the predicted class. The fact that the train and validation values are similar denotes that the model has a fairly good generalization ability. Figure 3 shows how these values evolved over time, through the 100 epochs of training.

Table 2. Main performance metrics of the developed model.

Metric	Value	Loss	Value
mAP@[.5:.95]	0.59	Train Box Loss	0.026
mAP@0.5	0.96	Train Object Loss	0.0033
Precision@0.5	0.94	Train Class Loss	0.005
Recall@0.5	0.96	Validation Box Loss	0.084
		Validation Object Loss	0.0039
		Validation Class Loss	0.018

To further assess the generalization ability of the model, it was also evaluated on the test set, containing 36 images. Figure 4 (left) shows the resulting confusion matrix. The Color, Hole and Metal Contamination classes are accurately identified by the model.

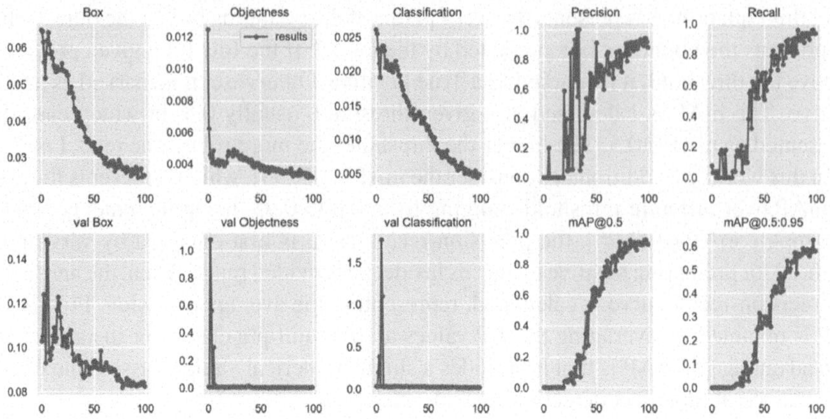

Fig. 3. Plots of box, objectness and classification loss (train and validation), and of precision, recall and mAP over the 100 epochs of training.

Cuts are confused with Holes 25% of the time. Indeed, from our observations, cuts are often marked as holes, and the defects are indeed similar and often difficult to distinguish even for a trained human (see for instance Fig. 1b and c). The confusion matrix also shows that 20% of Thread defects are False Negatives. This may be related to the size of some of these defects, which often contain very small and thin threads that are difficult to distinguish. Still, this improved when we moved from YOLOv3, which performed significantly worse in detecting such defects. Finally, 20% of the images containing no defects were classified as having a Metal Contamination defect. Once again, this may be due to the size of the metal pieces in this class, which are often very small, to the point of being almost confused with the threads of the fabric (as shown in Fig. 1d).

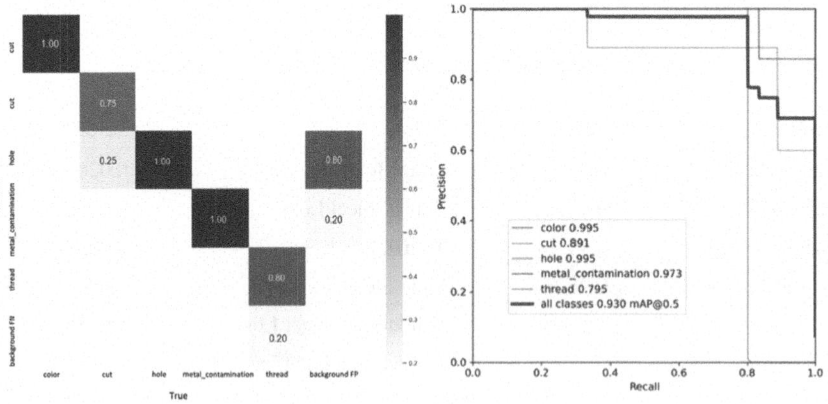

Fig. 4. Confusion Matrix for the developed model, calculated on the test data (left) and Precision-recall curve for the 5 object classes and considering all classes, calculated using a IoU threshold of 0.5 on the test data (right).

The precision-recall curve for each class and for all classes, for a threshold of 0.5, is also shown in Fig. 4 (right). Examples of detections (one for each class) obtained from the trained model and the respective degree of certainty are provided in Fig. 5.

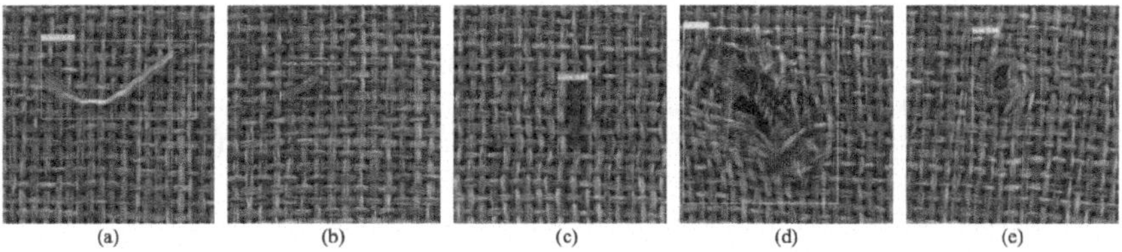

Fig. 5. Five examples of detections (zoomed in) and corresponding certainties obtained by the developed model on the test data, left to right: (a) thread (0.92), (b) metal contamination (0.59), (c) color (0.81), (d) cut (0.86) and (e) hole (0.66).

4 Conclusions

The textile industry, like many others, is seeking ways to automate and improve the performance of quality inspection systems. Computer vision techniques currently play a pivotal role in this regard. In this paper we investigated the hypothesis of using a model of the YOLO family to perform defect detection and localization. Results show that it is possible to do it with relatively high accuracy, while being able to obtain the characteristics of the defect, such as location on the fabric, height, width or area. These are paramount for future phases of this project, in which open challenges such as the relationship between production factors and prevalence of defects and their characteristics will be addressed, with the aim of minimizing defects when setting up production. Future work will be supported by the ongoing installation of cameras for collecting images from Jacquard looms in real-time, so that a more extensive dataset can be built and, ultimately, implement a defect detection system. The main open research challenge is to ascertain whether this architecture is good enough to, given the necessary training samples, generalize to previously unseen fabrics/patterns, or whether specific models will have to be trained.

Acknowledgements. This work is co-financed by Component 5 - Capitalization and Business Innovation, integrated in the Resilience Dimension of the Recovery and Resilience Plan within the scope of the Recovery and Resilience Mechanism (MRR) of the European Union (EU), framed in the Next Generation EU, for the period 2021–2026, within project Produtech_R3, with reference 60.

References

1. Bergmann, P., Batzner, K., Fauser, M., Sattlegger, D., Steger, C.: The MVTec anomaly detection dataset: a comprehensive real-world dataset for unsupervised anomaly detection. Int. J. Comput. Vision **129**(4), 1038–1059 (2021)
2. Terven, J., Cordova-Esparza, D.: A comprehensive review of YOLO: from YOLOv1 to YOLOv8 and beyond. arXiv preprint arXiv:2304.00501 (2023)
3. Wang, C.-Y., Bochkovskiy, A., Liao, H.-Y.M.: YOLOv7: trainable bag-of-freebies sets new state-of-the-art for real-time object detectors. In: Proceedings of the IEEE/CVF Conference on Computer Vision and Pattern Recognition, pp. 7464–7475 (2023)
4. Everingham, M., et al.: The Pascal Visual Object Classes (VOC) challenge. Int. J. Comput. Vision **88**, 303–338 (2010)
5. Athira, M.V., Khan, D.M.: Recent trends on object detection and image classification: a review. In: 2020 International Conference on Computational Performance Evaluation (ComPE), pp. 427–435. IEEE (2020)
6. Azamfirei, V., Granlund, A., Lagrosen, Y.: Multi-layer quality inspection system framework for industry 4.0. Int. J. Autom. Technol. **15**(5), 641–650 (2021)
7. Redmon, J., Divvala, S., Girshick, R., Farhadi, A.: You only look once: unified, real-time object detection. In: Proceedings of the IEEE Conference on Computer Vision and Pattern Recognition, pp. 779–788 (2016)
8. Ghiasi, G., Lin, T.Y., Le, Q.V.: DropBlock: a regularization method for convolutional networks. In: Advances in Neural Information Processing Systems, vol. 31 (2018)
9. Rasheed, A., et al.: Fabric defect detection using computer vision techniques: a comprehensive review. Math. Probl. Eng. **2020**, 1–24 (2020)
10. Li, C., Li, J., Li, Y., He, L., Fu, X., Chen, J.: Fabric defect detection in textile manufacturing: a survey of the state of the art. Secur. Commun. Netw. **2021**, 1–13 (2021)

Vision-Based Ladle Monitoring System for Steel Factories

Mohamed Selim[1(✉)], Pablo López de Uralde[2], Jon Mata[2], Eider Gorostegui-Colinas[2], Beatriz Chicote[2], Alain Pagani[1], and Didier Stricker[1]

[1] German Research Center for Artificial Intelligence, DFKI GmbH, Kaiserslautern, Germany
mohamed.selim@dfki.de
[2] Lortek, Ordizia, Spain

Abstract. This paper presents a vision-based ladle monitoring system for steel factories, consisting of two modules: one for ladle surface temperature analysis using thermal cameras and another for deep learning-based detection and recognition of ladle identification numbers. The first module monitors ladle thermal behavior by capturing high-resolution thermal images and employing advanced image analysis techniques. This enhances safety and efficiency in steel production. The second module focuses on digit recognition on the ladle surface, providing crucial identification and tracking information. A robust deep learning model trained on a large dataset of thermal camera images achieves high accuracy in ladle identification. The proposed system integrates thermal cameras and advanced image analysis techniques, offering real-time monitoring, early anomaly detection, and accurate ladle identification. Experimental evaluations demonstrate its effectiveness, indicating its potential for practical implementation in steel factory environments.

Keywords: Ladle monitoring · Thermal cameras · Deep Learning · Steel manufacturing

1 Introduction

A ladle is a steel vessel used for transporting liquid steel in a factory. Efficient coordination of ladle movements is crucial to minimize heat losses and ensure optimal supply. The lifetime of a ladle depends on refractory lining wear, which can have significant economic and safety impacts if not addressed promptly. Premature retirement of ladles increases refractory costs and overall production expenses. Currently, ladle lifetime decisions rely on visual observations by experienced personnel. However, existing commercial laser measuring systems lack precision in assessing the remaining refractory layer, mainly focusing on major cracks. Quality monitoring using cameras in the industry has revolutionized manufacturing processes, enabling real-time visual inspections that enhance product consistency and minimize defects [5,6]. These advanced camera systems not only detect imperfections but also provide valuable data for process optimization,

A. Wagner et al. (Eds.): ESAIM 2023, LNME, pp. 185–194, 2024.
https://doi.org/10.1007/978-3-031-57496-2_19

ensuring higher standards of quality control. Thermal cameras offer the ability to monitor ladle surface temperature, detect refractory wear-related issues, and track temperature changes. This allows for the identification of hotspots near breakout risks and the prediction of future ladle behavior by analyzing temperature evolution along different profiles. As ladle parts experience varying thermal history and wear, thermal monitoring provides a comprehensive evaluation of both current and future ladle states.

Considering the cycle of a ladle from the Electric Arc Furnace (EAF) to the casting station, the ladle is tracked and identified by the factory system. However, tracking and identification are not accurately performed at the cleaning and preheating stations. As a result, the time spent by the ladles at these stations is not precisely calculated. This time information is valuable for planning ladle usage in heats. If a ladle spends a short time after cleaning at the burners, it requires less preheating before it can be used in the next heat. Therefore, ladle identification at these stations is important. Each ladle in the steel factory has a specific number written on its body using steel pieces. The ladle number is welded on the surface, typically in multiple locations. Computer vision techniques can be employed to locate and identify the ladle number accurately. However, the harsh environment at the steel factory, including dust, can obstruct vision. Therefore, careful consideration and study are required when choosing cameras for this task. Color and thermal images have been explored for ladle tracking, each with its own advantages and disadvantages compared to the other.

The innovation of the proposed method lies in its comprehensive approach to ladle monitoring within steel factories. By combining thermal cameras and advanced image analysis techniques, this system not only tracks ladle surface temperatures with precision but also achieves digit recognition for ladle identification numbers through deep learning. This integrated approach enhances safety and efficiency in steel production, offering real-time monitoring, early anomaly detection, and accurate ladle identification, ultimately paving the way for practical implementation in industrial environments.

2 Ladle Surface Monitoring

The measurements for working on the thermal characterization of the ladles were made with a Flir a655sc microbolometric camera [7]. Microbolometer FPAs can be created from metal or semiconductor materials and operate according to non-quantum principles. This means that they respond to radiant energy in a way that causes a change of state in the bulk material (i.e., the bolometer effect). Generally, microbolometers do not require cooling, which allows compact camera designs that are relatively low in cost [8]. Thermal images were captured at the billet casting machine turret position. The turret holds ladles for casting and allows quick ladle changes using its two arms. While one ladle is being cast, an empty ladle can be loaded from the opposite side. The thermal camera measures the ladle in this opposite position. During casting, the ladle is filled with hot steel, and after emptying, it appears empty in the camera. The system consists of acquisition, data extraction, and data analysis nodes.

The acquisition node captures data from the thermal camera and filters the sequences to store only necessary and non-redundant information. It monitors a specific directory where the camera saves thermal sequences and related data in a format specific to the camera and software (e.g., ".seq" files generated by ResearchIR software from FLIR). Whenever a new file is added to the directory, the acquisition node algorithm processes it. The algorithm converts the raw data of each sequence into a 3D array, F(i,j,t), with dimensions determined by the camera's pixel resolution and the number of frames captured over time. This stage selects the optimal frame for post-processing based on two criteria. First, the frame must have a maximum temperature exceeding 400 °C to ensure a ladle is present in the camera's field of view. Second, the chosen frame minimizes the absolute mean optical flow per frame. This criterion accounts for the stationary period of the ladle after the crane positions it in the tower, as well as its rest period before being lifted again after the casting process.

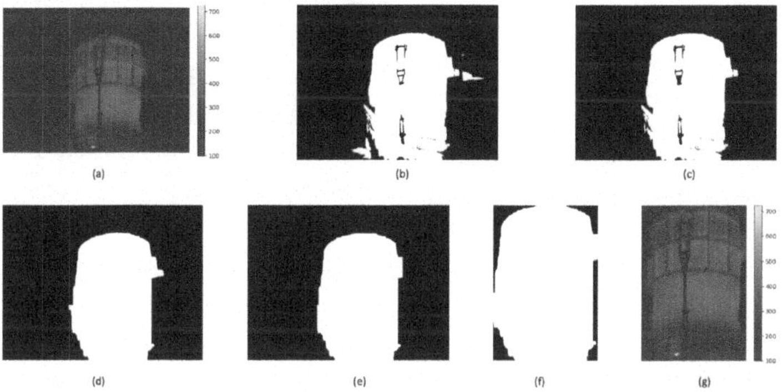

Fig. 1. Ladle recognition algorithm

After the selection of the most appropriate frame, a pre-processing has been developed to eliminate areas that are not of interest for the posterior analysis, such as background, reflections or defective pixels. The steps taken are described below and the effect of each step can be appreciated in Fig. 1. The algorithm takes the frame selected on previous paragraph. Every pixel with a temperature value below of 150 C$^{\underline{o}}$ is discarded. The greatest continuous area is detected, and every other area is discarded. Afterwards, some morphological transformations are applied. A 50 iterations dilation with a 5×5 pixel is followed by an erosion of the same characteristics. Every pixel is discarded if the non-masked pixel amount in its row or column does not surpass the thresholding value of 60 pixels. The mask is cropped so that only the relevant area of the image is saved. The output image is resized with the same dimensions as the mask.

Thanks to this pre-processing, the ladle is centred in the stored image, and it is ready be sent to the data extraction node for obtaining the information within the image and posterior data analysis. Besides this, the total amount of information transmitted between nodes is reduced to the minimum necessary, optimizing times in the processing. With this stage, the acquisition node concludes its work, reading a file in a specific format, and returning the resized image, mask and the metadata stored in the recording file (Fig. 2).

The objective of the data extraction node is to extract data from the input provided by the acquisition node for posterior data analysis. The extracted data is stored within the image, but for communication purposes, it is necessary to reduce the data amount to avoid saturating the architecture. The refractory degradation depends on the material it comes in contact with. Three zones can be distinguished in a ladle: the air zone, the slag zone, and the barrel zone. The ladle's nerves are used to differentiate these zones. The algorithm then identifies the hottest spots in each zone. The information from the acquisition node and the ladle's life cycle is combined and stored in a JSON file, significantly reducing the data size. To establish a unified coordinate system, a center is determined for each image, which serves as a common reference point. Artificial vision techniques are employed to locate the nerves or interfaces between the different zones. Finally, the extracted information, including data from the sequence file, caster-provided data, segmentation details, and thermal data, is stored in the JSON file, including temperature values, spot localization, and other relevant information (Fig. 3).

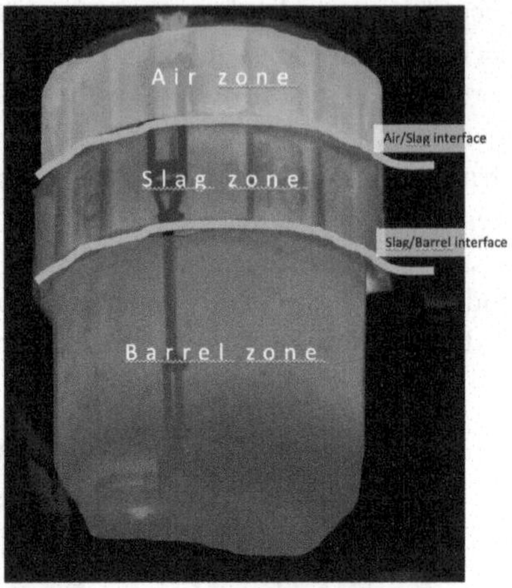

Fig. 2. Graphical description of the ladle's zones

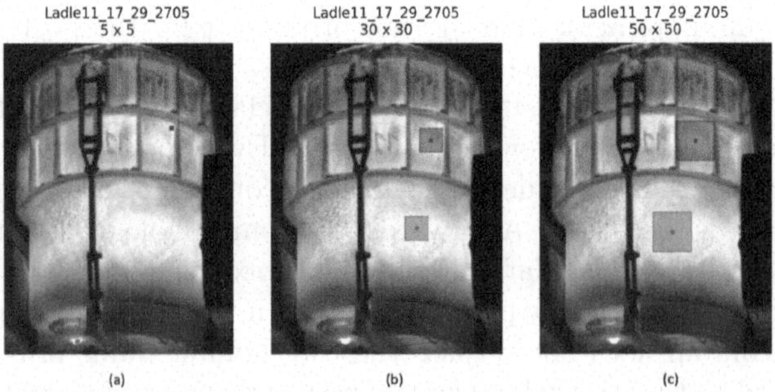

Fig. 3. Different size hottest areas for unique measurement in the slag zone (red) and in the barrel zone (blue). (a) 5×5 area, (b) 30×30 area and (c) 50×50 area (Color figure online)

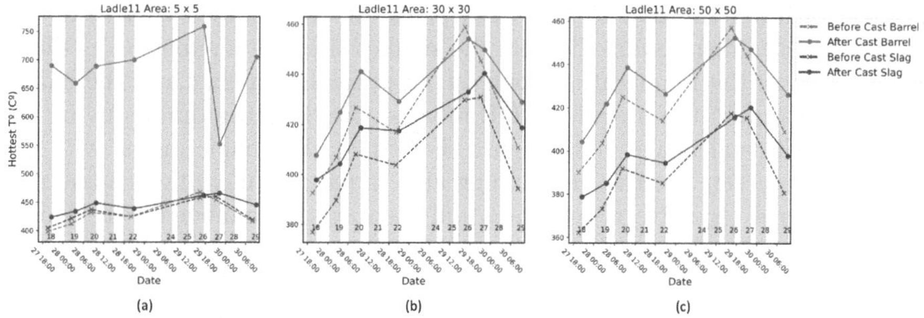

Fig. 4. Ladle 11 temperature analysis for areas (a) 5×5, (b) 30×30 and (c) 50×50

Observing the 5×5 chart (Fig. 4a), the effect of molten steel in the bottom of the ladle (barrel zone) is evident, with noticeably higher temperature values compared to other curves. However, there is an anomaly in the curve (use 27) where the temperature measured was lower but still influenced by molten steel. Analyzing the other three curves in Fig. 9a, two main conclusions can be drawn. First, the slag zone slightly heats up during the casting process, as indicated by higher after-casting temperatures. Second, temperature values in the slag zone are vaguely higher than in the barrel zone (pre-casting), possibly due to more significant slag erosion compared to molten steel. Figures 4 a–c confirm that temperatures after casting are slightly higher. Vertical profiles can provide more information about the casting process than individual temperature spots.

When considering larger areas (Fig. 4b–c), the previously mentioned patterns are not visible. Barrel temperatures are higher than those in the slag zone, with a greater difference in larger areas. Two reasons explain this: first, the larger areas partially cover the colder nerves, which are colder than the rest of the ladle surface. Second, temperature distribution is more homogeneous in the barrel zone, while the slag zone has hot spots closer to colder areas, leading to counterproductive effects on the overall mean temperature of the hottest area.

3 Ladle Tracking

Although several methods [1–3,9] and datasets [4] were introduced for number detection, they are not suitable to be used in steel factories domain because of the image appearance difference. In steel factories, the environment is dark, dusty, and usually very hot. Color cameras typically have higher resolution compared to thermal cameras. Initially, the detection algorithm was developed using thermal images, and later sample color images were examined. The color images revealed challenges in detecting the ladle number due to low contrast with the ladle surface, especially in a harsh, dusty environment. Additionally, using RGB color cameras raises concerns about worker security and data privacy in the steel factory setting. Investigating the thermal images revealed the following observations. The resolution of the thermal camera is smaller compared to color cameras, with a resolution of 388×288 pixels in the installed camera at the steel factory cleaning station. Despite the smaller size, the numbers can still be detected due to the temperature contrast between the number and the ladle's surface. In contrast, the provided RGB images pose difficulty in identifying the ladles. One thermal image in the figure shows two identifiable digits, with the digit shown in green and the surrounding area in red or a mix of green and red. The algorithm's objective is to locate and identify the number on the ladle's surface.

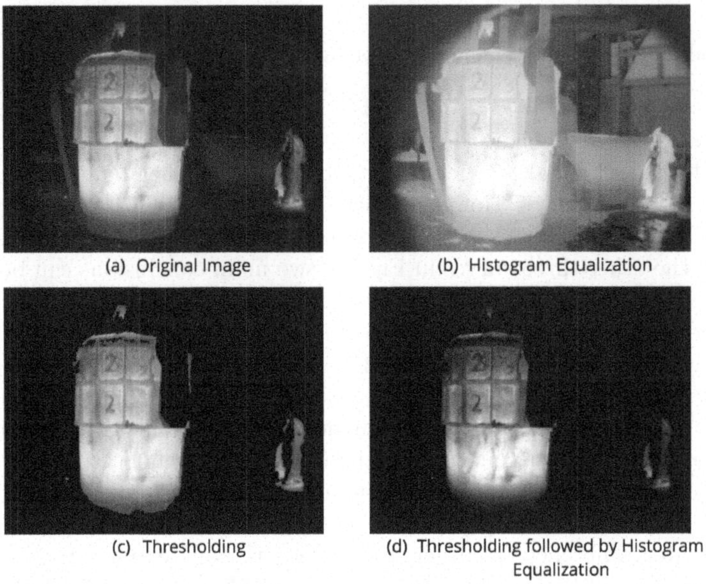

Fig. 5. Thermal Data preprocessing pipeline.

The ladle is the main object of interest in the camera's view and the hottest spot. The ladle's number is welded on its surface at least once, and the contrast difference between the ladle's surface and the number is important. Image pre-processing techniques were applied to enhance this contrast. Figure 5 shows the results of different pre-processing techniques applied to thermal images. Figure 5 illustrates the image pre-processing techniques used on the thermal data. The original heat map is shown in subfigure (a), which is represented in grayscale as a one-channel image to capture heat variation. Histogram equalization (subfigure (b)) enhances details in the image, revealing background information of the cleaning station that was not visible in the original image. To eliminate this background data, thresholding is applied (subfigure (c)), squashing the corresponding pixels to zero. The final image is obtained by applying histogram equalization to the thresholded image, enhancing the contrast between the ladle's number and the surrounding (colder) pixels. This contrast enhancement is crucial for ladle number identification. At the cleaning station, a thermal camera is installed with an adjusted perspective to acquire a better view of the ladles. The output of the camera is a two-dimensional array containing pixel temperatures. The initial data provided for the project was from the old perspective, which was sufficient for initial experiments but not suitable for model validation. The thermal data is a one-channel, two-dimensional image, and generating a heat map results in a three-channel RGB image, regardless of the color scales used (Fig. 6).

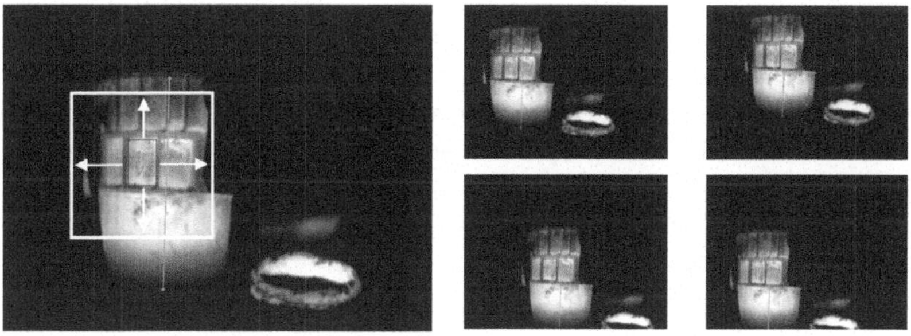

Fig. 6. Synthetic data generation pipeline

Thermal data consisting of approximately 20,000 images covering a period of around 3 months was shared. Each image was checked, and those containing a visible number were annotated with the number's bounding box and class (e.g., class 2 for ladle number 2). This annotation process resulted in a dataset of approximately 1,200 images. However, the data had an imbalance issue. Ladle 2 appeared approximately 700 times, while ladle 12 appeared only 49 times. To address this, synthetic data was generated to validate the developed method. The synthesized images were created to ensure a balanced representation of all classes. The specific ladle numbers used in the steel factory (ladles 1, 2, 8, 10, 11, 12, 14, 15, 16, and 17) were taken into account during the data synthesis process (Fig. 7).

Twenty seven images were selected to be used as background images. The 27 images were annotated with a location to add the digit at, and a line consisting of two points to represent the orientation of the ladle. The angle of the line can be computed and used to rotate the number before resizing and adding it at the bounding box location. Since the number is colder than the surrounding surface. The intensity of the number was set to the reduced value of the average intensity of the bounding box. Having 27 images would provide a good variation in the background.

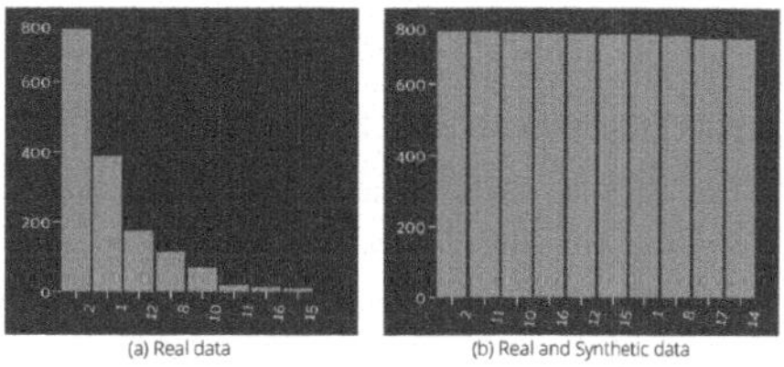

Fig. 7. Ladle images histograms

However, it is not sufficient to generate hundreds of images to try to balance the real data with the synthetic ones. Consequently, image translation was applied on the whole background image.

The ladle tracking task can be decomposed into two main problems, number detection in the thermal image, and number classification of the detected number. In the proposed solution, the numbers are modelled as object, thus making it an object detection task. In the proposed solution, and since there are a limited number of ladles, each ladle number is considered a class. Consequently in the developed model, the data has 10 classes. There are several methods that exist in literature for object detection. One of the state of the are methods is the Faster RCNNs. This methods belongs to the region proposal-based methods for object detection. The method has two deep neural networks, one for each sub task. The first one is for proposing the region where the number could exist in. The second network is for the number classification. The Faster-RCNNs can be implemented using several networks as backbone. In the proposed implementation, Residual Networks ResNet-50 was employed. The training is done in a supervised learning-based fashion. The labelled data is used to train the networks. The annotations either from the real or the synthetic data are in the training.

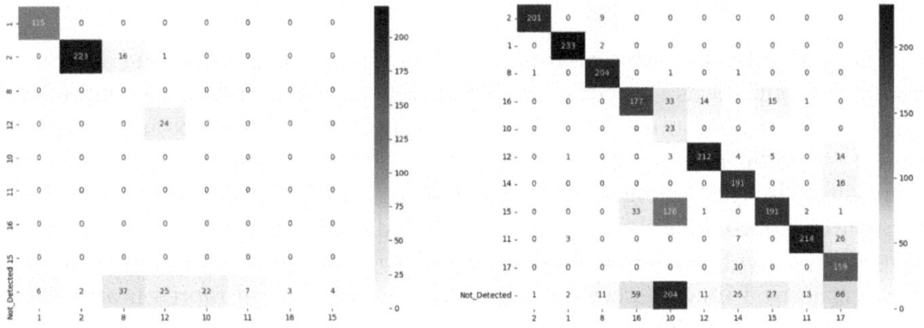

Fig. 8. Confusion Matrix (a) Real Data, (b) Real and Synthetic Data

Experiments and Results. Initial experiments were conducted using real data, which consisted of approximately 1200 annotated images. The dataset was divided into 70% for training and 30% for testing. Faster-RCNN Region-Proposal-Network was utilized with pre-trained weights. Training for 1000 epochs yielded satisfactory accuracy, considering the relatively small dataset. The experiment results, depicted in Fig. 8, present a confusion matrix plot of the model's detections on the test set. Ladles 1 and 2 were consistently identified correctly, but accuracy decreased for other classes due to insufficient training data. Consequently, as mentioned in the previous subsection, synthetic data was generated and the dataset was augmented. The dataset is balanced. In total, it contains 7100 image. Similar to the previous experiment, the training set used in this experiment was 70% of the data and the test set was 30%. With a significant increase in data volume compared to the previous experiment, the number of iterations was raised to 4000 epochs. The results are presented in Fig. 20, showing that almost all classes are accurately detected and identified. The confusion matrix highlights the False Negative detections in the last row. Additionally, an accuracy measurement was computed using a test set consisting of approximately 2200 real and synthetic images. A ladle is considered correctly detected if at least one number on the ladle is accurately detected, even if a second number is present but goes undetected. The proposed method achieved an accuracy of 80%, indicating that around 1760 images were correctly identified by the trained model. However, there appears to be an issue in detecting ladle 10, possibly due to the low resolution of the number in the image. Initial experiments were conducted using approximately 1200 annotated images, with a 70% training set and a 30% testing set. Faster-RCNN Region-Proposal-Network with pre-trained weights was employed and trained for 1000 epochs. Despite the relatively small dataset, satisfactory accuracy was achieved. The experiment results, shown in Fig. 8, display a confusion matrix plot of the model's detections on the test set. Ladles 1 and 2 were consistently identified correctly, but accuracy decreased for other classes due to insufficient training data. To address this, synthetic data was generated and augmented, resulting in a balanced dataset of approximately 7100 images. Sample detections can be seen in Fig. 21. The confusion matrix highlights the presence of False Negative detections in the last row. Additionally, an accuracy measurement using a test set of approximately 2200 real and synthetic images yielded an 80% accuracy, indicating that around 1760 images were correctly identified by the trained model. However, there appears to be an issue in detecting ladle 10, possibly due to the low resolution of the number in the image.

4 Conclusion

The work carried out during this study has quantified some information that steelmakers knew beforehand but were proofed thanks to the thermal camera. It was shown that, for the example given with ladle analysed in paragraph 3.3, the slag zone suffers greater erosion due to the chemical corrosion of the slag and therefore offers greater surface temperatures. Besides that, before and after casting differences were analysed and the temperature difference between both images was studied, proofing that the ladle surface remains heating while the casting process, as this process is long and by no means instantaneous.

In the ladle tracking it is shown that it is possible to detect and identify the ladles in the thermal images at the steel factory using computer vision and deep learning-based methods. The use of thermal images was employed, and favored over RGB images in such harsh environment. Data pre-processing was presented for reducing the unnecessary information in the image and enhancing the important section in the thermal image. Synthetic data was presented to balance the dataset. The underlying task was modelled as an object detection problem. The deep learning method used was the state-of-the-art Faster RCNN method.

Acknowledgment. This work has been funded by the EU project HyperCOG (Grant agreement number 869886).

References

1. Boonnag, C., et al.: PACMAN: a framework for pulse oximeter digit detection and reading in a low-resource setting. IEEE Internet Things J. **10**(15), 13196–13204 (2023)
2. Liu, H., Bhanu, B.: Pose-guided R-CNN for Jersey number recognition in sports. In: Proceedings of the IEEE/CVF Conference on Computer Vision and Pattern Recognition Workshops (2019)
3. Maglo, A., Orcesi, A., Pham, Q.C.: Efficient tracking of team sport players with few game-specific annotations. In: Proceedings of the IEEE/CVF Conference on Computer Vision and Pattern Recognition, pp. 3461–3471 (2022)
4. Netzer, Y., Wang, T., Coates, A., Bissacco, A., Wu, B., Ng, A.Y.: Reading digits in natural images with unsupervised feature learning (2011)
5. Papavasileiou, A., Aivaliotis, P., Aivaliotis, S., Makris, S.: An optical system for identifying and classifying defects of metal parts. Int. J. Comput. Integr. Manuf. **35**(3), 326–340 (2022)
6. Stavropoulos, P., Papacharalampopoulos, A., Sabatakakis, K.: Robust and secure quality monitoring for welding through platform-as-a-service: a resistance and submerged arc welding study. Machines **11**(2), 298 (2023)
7. Systems, F.: FLIR A655sc|thermal imaging camera|FLIR systems. https://www.flir.eu/products/a655sc. Accessed Aug 2023
8. Yadav, P.K., Yadav, I., Ajitha, B., Rajasekar, A., Gupta, S., Ashok Kumar Reddy, Y.: Advancements of uncooled infrared microbolometer materials: a review. Sens. Actuators A Phys. **342**, 113611 (2022). https://doi.org/10.1016/j.sna.2022.113611, https://www.sciencedirect.com/science/article/pii/S0924424722002497
9. Zhang, R., Wu, L., Yang, Y., Wu, W., Chen, Y., Xu, M.: Multi-camera multi-player tracking with deep player identification in sports video. Pattern Recogn. **102**, 107260 (2020)

Author Index

A. Wagner et al. (Eds.): ESAIM 2023, LNME, pp. 195–196, 2024.
https://doi.org/10.1007/978-3-031-57496-2

The manufacturer's authorised representative in the EU is Springer Nature Customer Service Centre GmbH, Europaplatz 3, 69115 Heidelberg, Germany. If you have any concerns regarding our products, please contact ProductSafety@springernature.com

Printed and bound by CPI Group (UK) Ltd, Croydon, CR0 4YY
01/10/2026
02225861-0010